SUBSTANCE ABUSE ASSESSMENT, INTERVENTIONS AND TREATMENT

METHAMPHETAMINE

METH LAB TRENDS AND INGREDIENT CONTROL EFFORTS

SUBSTANCE ABUSE ASSESSMENT, INTERVENTIONS AND TREATMENT

Additional books in this series can be found on Nova's website under the Series tab.

Additional E-books in this series can be found on Nova's website under the E-book tab.

ALCOHOL AND DRUG ABUSE

Additional books in this series can be found on Nova's website under the Series tab.

Additional E-books in this series can be found on Nova's website under the E-book tab.

SUBSTANCE ABUSE ASSESSMENT, INTERVENTIONS AND TREATMENT

METHAMPHETAMINE

METH LAB TRENDS AND INGREDIENT CONTROL EFFORTS

ACE M. DONIJAN
EDITOR

New York

For permission to use material from this book please contact us:
Telephone 631-231-7269; Fax 631-231-8175
Web Site: http://www.novapublishers.com

Additional color graphics may be available in the e-book version of this book.

Library of Congress Cataloging-in-Publication Data

ISBN: 978-1-62618-497-8

Published by Nova Science Publishers, Inc. † New York

CONTENTS

PREFACE

Methamphetamine (meth) is a powerful, highly addictive stimulant drug that has limited medical uses. Today, meth can be made by anyone using easily obtainable household goods and consumer products. According to the Drug Enforcement Administration (DEA) and the Office of National Drug Control Policy (ONDCP), illicit meth used in the United States today is manufactured by Mexican drug-trafficking organizations in "super labs" located in Mexico and California, as well as by cooks in "small toxic labs" predominantly located in the central United States, from the Gulf of Mexico to the Great Lakes and from the plains to the Appalachian Mountains. These labs pose significant public safety and health risks and financial burdens to local communities and states where they are found. The toxic chemicals and solvents involved in meth manufacturing can result in fiery explosions and expose property and people, including children, to contaminates that are dangerous and costly to remove. This book provides an overview on meth lab trends and ingredient control efforts relating to methamphetamine production.

Chapter 1 - Meth can be made by anyone using easily obtainable household goods and consumer products in labs, posing significant public safety and health risks and financial burdens to local communities and states where the labs are found. Meth cooks have discovered new, easier ways to make more potent meth that require the use of precursor chemicals such as PSE. Some states have implemented electronic tracking systems that can be used to track PSE sales and determine if individuals comply with legal PSE purchase limits. Two states, along with select localities in another state, have made products containing PSE available to consumers by prescription only. GAO was asked to review issues related to meth. Thus, GAO examined, among other things, (1) the trends in domestic meth lab incidents over the last decade; (2) the impact of electronic tracking systems on meth lab incidents and limitations of this approach, if any; and (3) the impact of prescription-only laws on meth lab incidents and any implications of adopting this approach for consumers and the health care system. GAO analyzed data such as data on meth lab incidents and PSE product sales and prescriptions. GAO also reviewed studies and drug threat assessments and interviewed state and local officials from six states that had implemented these approaches. These states were selected on the basis of the type of approach chosen, length of time the approach had been in use, and the number of meth lab incidents. The observations from these states are not generalizable, but provided insights on how the approaches worked in practice.

Chapter 2 - Methamphetamine is a highly addictive drug with potent central nervous system (CNS) stimulant properties. In the 1960s, methamphetamine pharmaceutical products were widely available and extensively diverted and abused. The placement of methamphetamine into schedule II of the Controlled Substance Act (CSA) in 1971 and the removal of methamphetamine injectable formulations from the United States market, combined with a better appreciation for its high abuse potential, led to a drastic reduction in the abuse of this drug. However, a resurgence of methamphetamine abuse occurred in the 1980s and it is currently considered a major drug of abuse. The widespread availability of methamphetamine today is largely fueled by illicit production in large and small clandestine laboratories throughout the United States and illegal production and importation from Mexico. In some areas of the country methamphetamine abuse has outpaced heroin and cocaine.

Chapter 3 – This is Methamphetamine Addiction: Progress, But Need to Remain Vigilant, by National Institute on Drug Abuse.

Chapter 4 – Methamphetamine is a central nervous system stimulant drug that is similar in structure to amphetamine. Due to its high potential for abuse, methamphetamine is classified as a Schedule II drug and is available only through a prescription that cannot be refilled. Although methamphetamine can be prescribed by a doctor, its medical uses are limited, and the doses that are prescribed are much lower than those typically abused. Most of the methamphetamine abused in this country comes from foreign or domestic superlabs, although it can also be made in small, illegal laboratories, where its production endangers the people in the labs, neighbors, and the environment.

In: Methamphetamine
Editor: Ace M. Donijan
ISBN: 978-1-62618-497-8

Chapter 1

Drug Control: State Approaches Taken to Control Access to Key Methamphetamine Ingredient Show Varied Impact on Domestic Drug Labs*

United States Government Accountability Office

Why GAO Did This Study

Meth can be made by anyone using easily obtainable household goods and consumer products in labs, posing significant public safety and health risks and financial burdens to local communities and states where the labs are found. Meth cooks have discovered new, easier ways to make more potent meth that require the use of precursor chemicals such as PSE. Some states have implemented electronic tracking systems that can be used to track PSE sales and determine if individuals comply with legal PSE purchase limits. Two states, along with select localities in another state, have made products containing PSE available to consumers by prescription only. GAO was asked to review issues related to meth. Thus, GAO examined, among other things, (1) the trends in domestic meth lab incidents over the last decade; (2) the impact of electronic tracking systems on meth lab incidents and limitations of this approach, if any; and (3) the impact of prescription-only laws on meth lab

* This is an edited, reformatted and augmented version of United States Government Accountability Office, Publication No. GAO-13-204, dated January 2013.

incidents and any implications of adopting this approach for consumers and the health care system. GAO analyzed data such as data on meth lab incidents and PSE product sales and prescriptions. GAO also reviewed studies and drug threat assessments and interviewed state and local officials from six states that had implemented these approaches. These states were selected on the basis of the type of approach chosen, length of time the approach had been in use, and the number of meth lab incidents. The observations from these states are not generalizable, but provided insights on how the approaches worked in practice.

WHAT GAO FOUND

Methamphetamine (meth) lab incidents—seizures of labs, dumpsites, chemicals, and glassware—declined following state and federal sales restrictions on pseudoephedrine (PSE), an ingredient commonly found in over-the-counter cold and allergy medications, but they rose again after changes to methods in acquiring PSE and in the methods to produce meth. According to Drug Enforcement Administration (DEA) data, the number of lab incidents nationwide declined through 2007 after the implementation of state and federal regulations on PSE product sales, which started in 2004. The number of meth lab incidents reported nationally increased after 2007, a trend primarily attributed to (1) the emergence of a new technique for smaller-scale production and (2) a new method called smurfing--a technique used to obtain large quantities of PSE by recruiting groups of individuals to purchase the legally allowable amount of PSE products at multiple stores that are then aggregated for meth production.

Electronic tracking systems help enforce PSE sales limits, but they have not reduced meth lab incidents and have limitations related to smurfing. By electronically automating and linking log-book information on PSE sales, these systems can block individuals from purchasing more than allowed by law. In addition, electronic tracking systems can help law enforcement investigate potential PSE diversion, find meth labs, and prosecute individuals. However, meth cooks have been able to limit the effectiveness of such systems as a means to reduce diversion through the practice of smurfing.

The prescription-only approach for PSE appears to have contributed to reductions in lab incidents with unclear impacts on consumers and limited impacts on the health care system. The implementation of prescription-only laws by Oregon and Mississippi was followed by declines in lab incidents. Law enforcement officials in Oregon and Mississippi attribute this reduction in large part to the prescription-only approach. Prescription-only status appears to have reduced overall demand for PSE products, but overall welfare impacts on consumers are unclear because of the lack of data, such as the cost of obtaining prescriptions. On the basis of the limited information available from health care providers in Oregon and Mississippi, there has not been a substantial increase in workload demands to provide and dispense prescriptions for PSE products.

ABBREVIATIONS

CHPA	Consumer Healthcare Products Association
CMEA	Combat Methamphetamine Epidemic Act of 2005
DD	Difference-in-Differences
DEA	Drug Enforcement Administration
DOJ	Department of Justice
EPIC	El Paso Intelligence Center
ETS	electronic tracking system
FDA	Food and Drug Administration
GEE	generalized estimating equation
HIDTA	High Intensity Drug Trafficking Area
ID	identification
LEOKA	Law Enforcement Officers Killed in Action
MDMA	Ecstasy (drug)
NDIC	National Drug Intelligence Center
NMPI	National Methamphetamine & Pharmaceuticals Initiative
NPLEx	National Precursor Log Exchange
N-SSATS	National Survey of Substance Abuse Treatment Services
NSS	National Seizure System
ONDCP	Office of National Drug Control Policy
OTC	over-the-counter
PDMP	prescription drug monitoring program
PE	phenylephrine
PSE	pseudoephedrine
QIC	quasi-likelihood indicator criteria
TEDS	Treatment Episode Data Set
UCR	Uniform Crime Report

January 31, 2013

The Honorable Dianne Feinstein
Chairman

The Honorable Charles Grassley
Co-Chairman
Caucus on International Narcotics Control
United States Senate

Methamphetamine (meth) is a powerful, highly addictive stimulant drug that has limited medical uses.[1] Today, meth can be made by anyone using easily obtainable household goods and consumer products. According to the Drug Enforcement Administration (DEA) and the Office of National Drug Control Policy (ONDCP), illicit meth used in the United States today is manufactured by Mexican drug-trafficking organizations in "super labs" located in Mexico and California, as well as by cooks in "small toxic labs" predominately located in the central United States, from the Gulf of Mexico to the Great Lakes and from the plains to the Appalachian Mountains.[2] These labs pose significant public safety and health risks and financial burdens to local communities and states where they are found. The toxic chemicals and solvents involved in meth manufacturing can result in fiery explosions and expose property and people, including children, to contaminates that are dangerous and costly to remove.

In the wake of an increased law enforcement focus on illicit meth labs in the 1980s, meth manufacturing methods changed. Meth cooks discovered new, easier ways to make more potent meth that required the use of precursor chemicals such as pseudoephedrine (PSE), a nasal decongestant commonly found in over-the-counter (OTC) cold and allergy medications. PSE is the primary essential ingredient used to make meth in the United States today.[3] Initial federal efforts to address a growing meth problem primarily focused on increased meth-trafficking penalties and regulating the bulk importation, exportation, and distribution of meth precursor chemicals such as PSE. However, by 2004, the annual number of meth lab incidents reported by law enforcement had reached an all-time high of over 24,000.[4] In response, many states began taking steps to further regulate PSE at the point of sale, such as requiring customers to present photo identification (ID) when purchasing PSE products and pharmacists to keep PSE products behind the counter and maintain a log of all sales. In 2006, the Combat Methamphetamine Epidemic Act of 2005 (CMEA) was passed to, among other things, regulate the retail sale of OTC PSE products by setting daily and monthly sales limits for customer purchases and requiring sellers to keep these products behind the counter and maintain a written or electronic sales logbook.[5]

Recognizing the serious public safety and health risks and financial burdens of these labs, state and local governments and Congress have since taken or considered taking further action to help prevent PSE diversion to make illicit meth. For example, some states have implemented electronic tracking systems that can be used to track PSE sales and determine if individuals comply with legal PSE purchase limits. Two states, along with select localities in another state, have made products containing PSE available to consumers by prescription only.[6] In addition, since 2010, at least 69 bills have been introduced in 18 states that would require consumers to obtain a prescription to purchase PSE products. In this context, you requested that we provide information on domestic meth lab incident trends and the impacts these two approaches have had on the domestic meth lab problem. Accordingly, this report addresses

- the trends in domestic meth lab incidents over the last decade and the impact of domestic meth labs on the communities affected by them;
- the impact of electronic tracking systems on domestic meth lab incidents and the limitations, if any, of using these systems; and
- the impact of prescription-only laws on domestic meth lab incidents and any implications of this approach for consumers and the health care system.

To identify trends in domestic meth lab incidents over the last decade, we analyzed data for all states from DEA's National Seizure System on lab seizure incidents that occurred during the last 10 complete calendar years, 2002 through 2011.[7] Using these data, we analyzed the number of incidents nationally, by region, and by type of lab and lab capacity. To assess the reliability of these data, we discussed the data with agency officials and compared them with other data and documentation, where available, from states we selected as case studies for this review. We selected this non-probability sample of states to reflect a mix of characteristics, such as the type of approach chosen for controlling the sale of PSE products (electronic tracking or prescription-only), length of time the approach has been in use, and the number of meth labs seized relative to the state's population size.[8] We worked with DEA to resolve any data discrepancies and determined that the data were sufficiently reliable for the purposes of this report. We also interviewed federal officials, as well as state and local officials in the 6 case study states about meth lab trends. While we cannot generalize any findings or results to the national level from this sample, this information provided perspectives on the factors affecting meth lab trends. We also reviewed drug threat assessments and reports by the National Drug Intelligence Center (NDIC) and information from DEA and ONDCP officials. We reviewed the methodology of the assessments and reports and found them sufficiently reliable for our purposes. To determine the impact of domestic meth lab incidents on the communities affected by them, we reviewed documentation from the Department of Justice (DOJ), data and documentation from DEA, a report from the DOJ Inspector General on DEA's meth lab cleanup program, reports from the RAND Corporation, and information from various state and local officials from our case study states.

To determine the impact of electronic tracking systems on domestic meth lab incidents, we analyzed DEA data from 2002 through 2011 on the number of meth lab incidents that were reported in the 3 states that have implemented electronic tracking the longest—Kentucky, Oklahoma, and Tennessee. We also reviewed PSE purchase activity data for 2011 and 2012 from Appriss, the firm that developed and manages the software program MethCheck, which is used as the operational platform for the National Precursor Log Exchange (NPLEx), the interstate electronic tracking system paid for by manufacturers of PSE products. We chose this time period because 2011 and 2012 were the most recent years for which data from multiple states were available. In addition, we obtained information from officials with Appriss as well as officials with state and local law enforcement in the 4 electronic tracking case study states. Although the perspectives of these state and local officials cannot be generalized across the broader population of state and local law enforcement agencies in electronic tracking states, their perspectives provided insights into and information on the use and impact of the approach in practice and its limitations.

To determine the impact of prescription-only laws on domestic meth lab incidents and any implications of adopting this approach for consumers and the health care system, we analyzed DEA data on meth lab incidents from Mississippi and Oregon and their border states from 2002 through 2011.[9] We also conducted a statistical modeling analysis of Oregon lab incident data but could not conduct a similar analysis for Mississippi because sufficient data were not yet available. We interviewed state and local officials in prescription-only states about the impact of these laws on lab incidents, and although their perspectives cannot be generalized across states, they provided insight for this report. We also reviewed PSE purchase data from neighboring states. For Mississippi specifically, we obtained data from IMS Health Inc. through DEA on the volume of PSE and phenylephrine sales for the period when the prescription-only approach was in effect for part of the year to the 2009 period, when it was not.[10] We also obtained and reviewed data provided by the Mississippi Board of Pharmacy's Prescription Drug Monitoring Program (PDMP) to assess the impact on consumers. We also reviewed a report prepared for the Consumer Healthcare Products Association (CHPA) and interviewed the state boards of pharmacy and state associations representing pharmacists in Mississippi and Oregon and the National Consumers League and the Asthma and Allergy Foundation of America.[11] We selected these organizations because they have previously surveyed consumers about access to PSE products. In addition, we interviewed state associations representing physicians practicing in Oregon and Mississippi and Medicaid program officials in Mississippi and Oregon. While their perspectives cannot be generalized to the larger population of physicians in these states, they provided insights into the impact of the approach on their members' practices. We assessed the reliability of data received from the NPLEx system, IMSHealth Inc., and the Mississippi PDMP by interviewing knowledgeable officials and reviewing relevant documentation, and we determined that these data were sufficiently reliable for the purposes of this report.

We conducted this performance audit from February 2012 to January 2013 in accordance with generally accepted government auditing standards. Those standards require that we plan and perform the audit to obtain sufficient, appropriate evidence to provide a reasonable basis for our findings and conclusions based on our audit objectives. We believe that the evidence obtained provides a reasonable basis for our findings and conclusions based on our audit objectives.

Additional details on our scope and methodology are contained in appendix I.

BACKGROUND

PSE—An Essential Ingredient for Making Meth

Meth is relatively easy and cheap to make today by individuals with little knowledge of chemistry or laboratory skills or equipment. PSE, an ingredient used in OTC and prescription cold and allergy medications, is the key substance needed to make the dextrorotatory methamphetamine (d-meth) illicitly produced in most domestic meth labs today. The difference between a PSE molecule and a d-meth molecule is a single oxygen atom. Meth cooks make d-meth by using common household products to remove this oxygen atom to produce meth as shown in figure 1.

PSE Uses, Restrictions, and Warnings

As a decongestant, PSE can provide temporary relief of sinus congestion by decreasing inflammation in the sinus cavities and ear canal. In 1976, the FDA found PSE to be safe and effective for use as an OTC product, and therefore PSE became marketable without prior approval from FDA under their monograph system. Shortly thereafter, PSE became marketed as an OTC through brand names such as Sudafed. According to CHPA, as of June 2012, there were about 40 different OTC products under 20 brand names containing PSE in their formulations. No children's medications containing PSE are currently being produced. The warnings section of the FDA-approved OTC PSE label instructs individuals with heart disease, high blood pressure, diabetes, and a few other conditions to consult a doctor or health professional before using. Manufacturing labels typically instruct individuals to discontinue use if symptoms persist longer than seven days.

Figure 1. Conversion of PSE to Meth.

Meth cooks have used two primary processes known as the Nazi/Birch and Red P methods to make d-meth.[12] In recent years, meth cooks have developed a variation of the Nazi/Birch method known as the One Pot or Shake and Bake method that produces meth in one step where ingredients are mixed together in a container such as a 2-liter plastic bottle. Another process for making meth is known as the P-2-P method, which produces a less potent form of meth known as racemic or dl-meth.[13] It does not require PSE as a precursor chemical and is typically half as potent as the d-meth made with PSE.

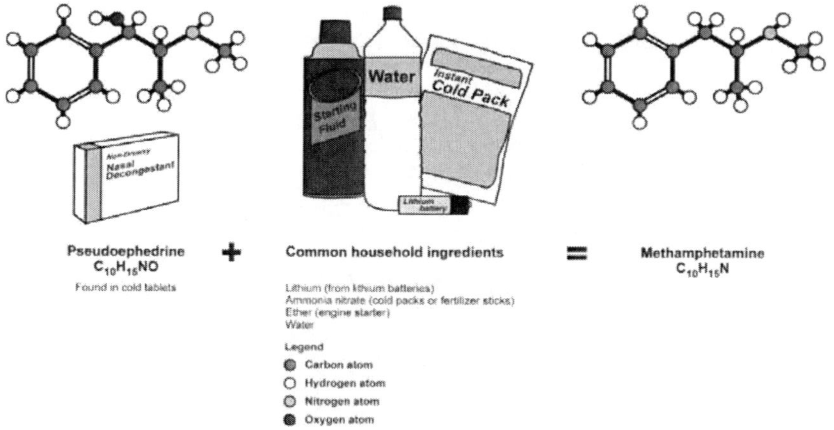

Source: GAO.

Figure 1. Conversion of PSE to Meth.

Federal and State PSE Sales Restrictions

Initial federal efforts to address a growing meth lab and abuse problem primarily focused on increasing meth-trafficking penalties and regulating the bulk importation, exportation, and distribution of meth precursor chemicals such as PSE.

In 2004, Oklahoma was the first state to pass a law to control the retail sale of PSE products by requiring customers to present photo IDs and pharmacists to keep the product behind the counter and log all sales. By November 2005, over 30 other states had passed laws related to the control of the retail sale of PSE products. In 2006, the CMEA was enacted, which included measures designed to control the availability of meth precursor chemicals such as PSE by regulating the retail sale of OTC products containing these chemicals.[14]

The CMEA placed restrictions on the sale of these products, including (1) requiring these products to be kept behind the counter or in a locked cabinet where customers do not have direct access; (2) setting a daily sales limit of 3.6 grams and a monthly purchase limit of 9 grams per customer regardless of the number of transactions; and (3) requiring sellers to maintain a logbook, written or electronic, to record sales of these products and verify the identity of purchasers.[15] The CMEA does not prohibit states from taking actions to establish stricter sales limits or further regulate the sale of PSE products.[16]

PSE Locking Formulations

In addition to the electronic tracking and prescription-only approaches, another potential approach to prevent the diversion of OTC PSE medications is to distribute the medications using a delivery system or formulation that precludes the extraction of the PSE from those medications. Under the CMEA, the Attorney General may exempt products that have been determined to be unusable in the illicit manufacture of methamphetamine from the CMEA's requirements (such as placement behind the pharmacy counter, enforcing purchase limits, and requiring identification and entries into a logbook of all PSE sales). According to DEA, at least one private company has asked DEA to determine if its delivery system meets the exemption criteria. DEA officials stated that while initial tests of the system have shown some promise, additional evaluation is required to determine whether the exemption criteria have been met. To date, no product has received an exemption.

Electronic Meth Precursor Tracking Systems and Prescription-Only Laws and Ordinances

Since the passage of the CMEA, some states have implemented electronic systems to track sales of products containing PSE. Through these systems, retailers report sales of PSE products to a centralized database that can be used to determine whether individuals are exceeding the purchase limitations of the CMEA or state laws. Reported information typically includes the date and grams purchased, as well as the name, address, and other identifying information of the purchaser. Most tracking systems are stop sale systems that would query the database, notify the retailer whether the pending sale would violate federal or state purchase limitations, and deny sales where limits have already been reached.[17] As of December 2012, 19 states were using stop sale tracking systems. Seventeen of these states were using a system called the NPLEx that is endorsed and funded by PSE manufacturers through CHPA.[18] Two states were using systems developed in-house or by another vendor.

Some states and localities have taken additional steps to regulate PSE sales. Oregon, Mississippi, and 63 Missouri cities or counties have passed laws or ordinances requiring individuals to obtain a prescription from a health care provider in order to purchase PSE products. While a prescription is required, an in-person encounter with a health care provider may not be necessary to obtain the prescription. There is no set limit to how much PSE can be prescribed. Both Oregon and Mississippi require that prescriptions for PSE products be entered into the states' prescription drug monitoring program, a program that allows for pharmacists and prescribers to electronically look up how much PSE product has been prescribed to a patient. Figure 2 shows the states with prescription-only laws and ordinances and electronic tracking systems, including the dates these systems were implemented.

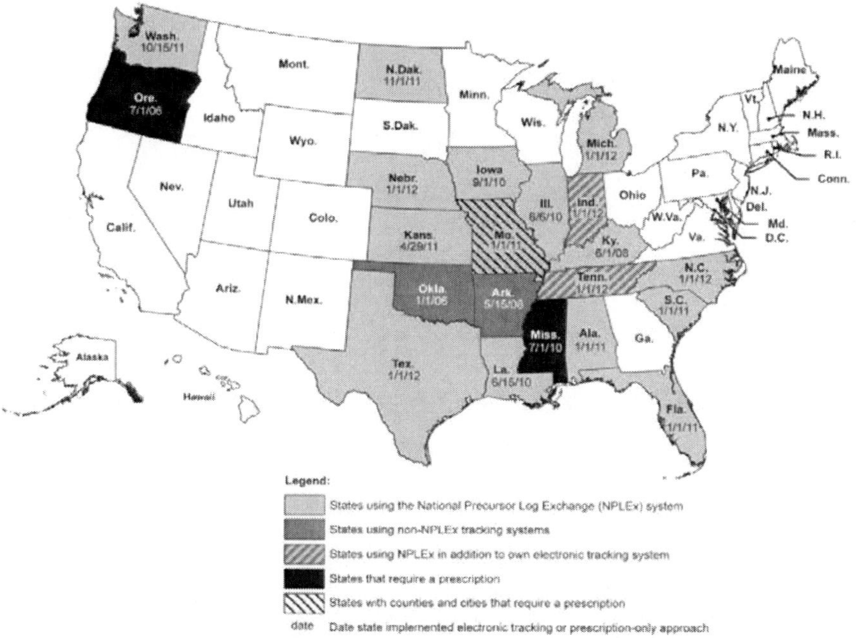

Source: GAO analysis of state statutes and information provided by the National Precursor Log Exchange administrator and the National Alliance for Model State Drug Laws; Map Resources (map).

Note: Analysis as of December 2012. An additional 7 states were scheduled to begin using NPLEx on 01/01/13.

Figure 2. Adoption of Electronic Tracking and Prescription-only Approaches, by State.

METH LAB INCIDENTS ROSE AFTER A SHARP DECLINE AND HAVE WIDE-RANGING IMPACTS ON COMMUNITIES

Meth Lab Incidents Declined Sharply Following Implementation of State and Federal PSE Sales Restrictions

According to DEA data on meth lab incidents, after peaking in 2004, the number of lab incidents nationwide declined through 2007 after the implementation of state and federal regulations on PSE product sales. As shown in figure 3, the number of lab incidents peaked in 2004, with states reporting over 24,000 lab incidents nationally.

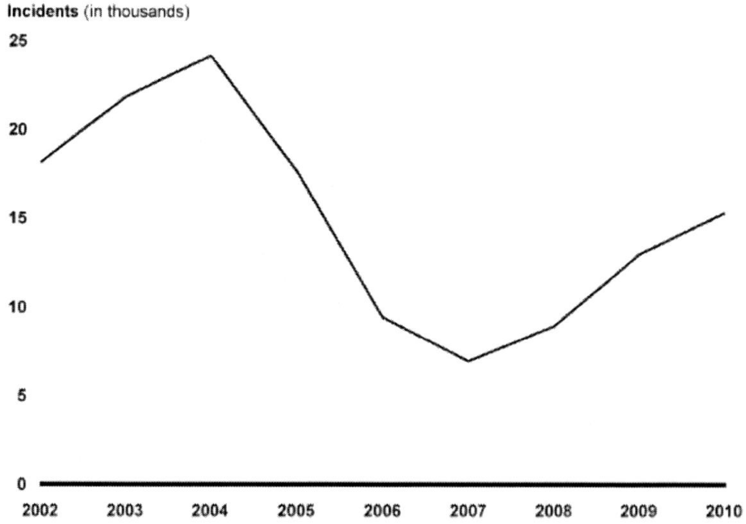

Year	2002	2003	2004	2005	2006	2007	2008	2009	2010
Incidents	18,131	21,817	24,155	17,866	9,426	6,951	8,933	12,971	15,314

Note: In 2011, law enforcement reported 13,530 lab incidents to DEA's National Seizure System (NSS). Law enforcement officials are required only to report incidents to the NSS in cases where DEA funds are used for cleanup. From February 2011, to September 2011, there were no DEA funds available for cleanup. Consequently, it is likely that the number of 2011 lab incidents was underreported.

Source: GAO analysis of data from DEA's National Seizure System.

Figure 3. Meth Lab Incidents Nationwide, 2002 through 2010.

However, beginning in 2005, the number of incidents began to decline sharply and reached a low of about 7,000 incidents in 2007. While there may be multiple factors at work that resulted in this decline such as region-specific factors, federal, state, and local law enforcement officials attribute the primary cause of the decline to the restrictions on purchases of PSE products imposed at both the federal and state levels from 2004 through 2006. The impact of these restrictions was to reduce the accessibility of PSE for use in illicit meth labs, which in turn resulted in fewer labs during this period.

Meth Lab Incidents Rose Following the Emergence of the One Pot Production Method and Smurfing

After reaching a low in 2007, the number of meth lab incidents reported nationally increased over the next few years. National trends show that meth lab incidents have increased since 2007, reaching more than 15,000 at the end of 2010 –more than double the number of reported incidents for 2007. Federal, state, and local law enforcement officials attribute this rising trend primarily to two factors:

- **The emergence of a new technique for smaller-scale production.** A production method popularly called the One Pot method, which simplified the entire meth production process down to a single 2 liter plastic bottle and enhanced the ability of individuals to make their own meth, began to emerge in 2007. With this method, meth addicts are capable of manufacturing their own meth quicker and with less PSE, chemicals, and equipment than required by traditional meth-manufacturing methods, although this method also produces less meth than the traditional manufacturing methods. According to DEA data, more than 87 percent (43,726) of the labs seized with a capacity reported from 2008 through 2011 have been smaller capacity (less than 2 ounce) labs and about 74 percent (39,049) used the Nazi/Birch manufacturing process, of which the One Pot method is a variation. Less than 0.5 percent (219) of the labs seized during this period were super labs (labs producing 10 pounds or more of meth per batch), less than 13 percent (6,473) used the Red P method, and only 0.05 percent (26) of the labs seized during this period used the P-2-P method, which does not require PSE as a precursor chemical.[19]

- **Use of a method for meth producers to circumvent PSE sales restrictions.** Another key factor federal, state, and local officials attribute to the increase in meth labs in recent years is the use of a method known as smurfing to work around PSE sales restrictions. Smurfing—which is discussed in greater detail later in this report—essentially involves the coordinated effort by individuals or groups of individuals to purchase the maximum per person legal allowable amount of PSE products and then aggregate their purchases for the use in meth production or for sale to a meth producer. Federal, state, and local officials stated that consequently, using this technique, meth producers have been able to obtain the PSE product they need to make meth despite the federal and state sales restrictions. This, in turn, has led to the proliferation of more labs.

South and Midwest Regions Have Significantly Increased Meth Lab Incidents Overall as Compared with Other Regions

Further examination of data trends at the regional level reveals that the number of meth lab incidents varies greatly among regions of the country. Specifically, while the number of meth lab incidents continues to be low in the Northeast and declines in the number of meth labs incidents have been maintained in the West since PSE sales restrictions went into place, the South and Midwest regions have experienced significant increases overall in the number of incidents since 2007. Further, the South and Midwest have also had more lab incidents than the West and Northeast since 2003 (see fig.4). In general, these trends are consistent across all categories of lab types and capacities, except for incidents involving the P-2-P labs and labs of larger capacities (10 pounds or greater), for which the West tended to report higher numbers of incidents overall.

Domestic Meth Labs Impact Communities in Matters Related to Health Care, Child Welfare, Law Enforcement Resources, and the Environment

Impacts on Health Care

Meth labs can have a significant impact on a community's health care system when labs catch on fire or explode, causing serious injuries and burns to meth cooks and other individuals that require costly medical treatment. Mixing chemicals in meth labs creates substantial risks of explosions, fires, chemical burns, and toxic fume inhalation. These burns and related injuries resulting from these events can be more serious than burns and injuries sustained through non-meth-lab-related causes. For example, a 2008 study conducted of meth and non-meth burn patients that received treatment in one hospital burn unit in Kalamazoo, Michigan, from 2001 through 2005, found that meth lab patients tended to have more frequent inhalation injuries, needed greater initial fluid resuscitation volume, required intubation more frequently, and were more likely to have complications than non-meth patients.[20]

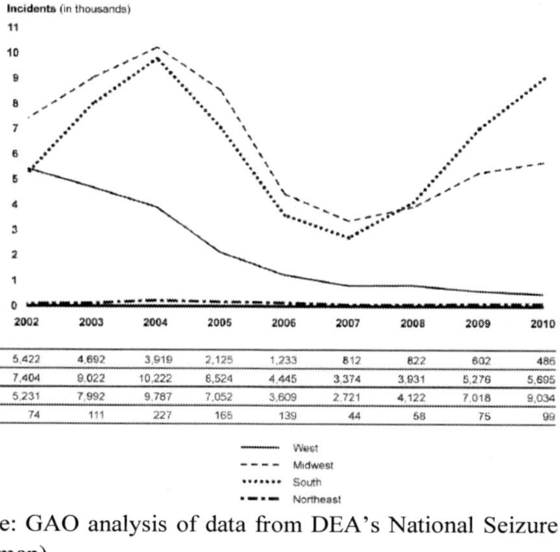

Year	2002	2003	2004	2005	2006	2007	2008	2009	2010
West	5,422	4,692	3,919	2,125	1,233	812	822	602	486
Midwest	7,404	9,022	10,222	8,524	4,445	3,374	3,931	5,276	5,695
South	5,231	7,992	9,787	7,052	3,609	2,721	4,122	7,018	9,034
Northeast	74	111	227	165	139	44	58	75	99

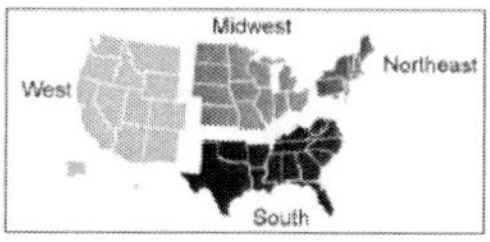

West
Midwest
South
Northeast

Source: GAO analysis of data from DEA's National Seizure System; Map Resources (map).

Notes:

Data was accessed on October 1, 2012.

Figure 4. Regional distribution of Meth Lab Incidents, January 2002 through 2010.

The small size of the relatively new One Pot or Shake and Bake method can be even more dangerous than larger meth labs, as drugmakers typically hold the One Pot container up close, increasing the risk for severe burns from the waist to the face. According to the director of the Vanderbilt University Regional Burn Center in Tennessee, meth lab injuries can also be more severe than burns resulting from just fires alone because patients often suffer thermal burns from the explosion, as well as chemical burns from exposure to caustic chemicals. He also noted that meth lab burn patients tend to be more difficult to treat because their addiction and overall poor physical health make it difficult for them to facilitate their own recovery as well as the fact that most attempt to hide the cause of their injury, which can hinder the administration of proper care.

The treatment for meth lab-related burns and injuries can be very expensive. According to one provider, treatment costs for two meth lab burn patients exceeded $2 million per patient. Although accurate estimates of the proportion of burn victims that received their burns from a meth lab are difficult[21] , one estimate placed the percentage of meth lab burn patients at 25 to 35 percent of total burn patients. Of those patients that are identified as receiving their injuries from meth labs, many are found either not to have health insurance or have publicly funded insurance such as Medicaid. For example, the 2008 Kalamazoo study also found that significantly fewer meth burn patients had private insurance, while more were on Medicaid or had no insurance as compared with non-meth burn patients.

Impacts on Children

Children who live at or visit locations or residences with meth labs or are present during drug production face acute risks to their health and safety. According to data from DEA's National Seizure System, over 21,000 children were reported affected by meth labs from 2002 through 2011. 22 Physically, the age-related behaviors of young children (such as frequent hand-to-mouth contact and physical contact with their environment) increase the likelihood that they will inhale, absorb, or ingest toxic chemicals, drugs, or contaminated food that may be within their reach or in their environment. For example, in 2009 a 20-month-old boy in Kentucky died from chemical burns to his trachea and bronchial system and toxic ingestion after accidentally drinking some liquid drain cleaner left over by adults who made meth in the trailer he was living in. A child living at a home with a meth lab may also inhale toxic substances or the secondhand smoke of adults who are using meth, receive an accidental skin prick from discarded needles or other drug paraphernalia, or absorb methamphetamine and other toxic substances through the skin following contact with contaminated surfaces, clothing, or food. The physiological characteristics of children, such as higher metabolic and respiratory rates and a developing central nervous system, also leave them vulnerable to the other effects of toxic chemical exposures, which can cause cancer; damage the brain, liver, kidney, spleen, and immunologic system; and result in birth defects.

In addition to the physical dangers, children in environments where meth is being made are also reported to be at risk to suffer abuse or neglect by their parents or other adults. Parents and caregivers who are meth dependent can become careless and often lose their capacity to take care of their children such as ensuring their children's safety and providing essential food, dental and medical care, and appropriate sleeping conditions. Children living in households where meth labs are operated are also at increased risk for being physically and sexually abused by members of their own family or other individuals at the site.

To protect the children discovered at meth lab sites from further harm and neglect, social service agencies remove the children from their homes and place them in foster care. Foster care is a social welfare service that serves the needs of abused and neglected children. Child welfare workers can remove a child if it is determined that remaining with the parents will jeopardize a child's welfare. Children are placed either with a surrogate foster family or in a residential treatment facility called a group home with the intent to provide temporary housing in a safe and stable environment until reunification with the child's birth parents or legal guardians is possible. Reunification happens once the state is convinced that the harmful factors that triggered removal no longer exist. Several states and jurisdictions have created special protocols and programs to address the needs of children exposed to clandestine meth labs. These protocols and programs typically involve medical screening of the children for toxicity and malnourishment, emergency and long-term foster care, and psychological treatment. Social service agencies may also seek to enroll meth-involved parents and their children in a family-based treatment program, where both the parents and children receive services. Family-based treatment programs offer treatment for adults with substance use disorders and support services for their dependent children in a supervised, safe environment that allows the family to remain together and prevents exposure to further harm. The costs to state department of human service agencies to provide services to these children can be significant depending on the number, age, and specific needs of the child. For example, from January 2006, through December 2011, the Missouri Department of Social Services substantiated 702 reports of children exposed to meth labs, involving a total of 1,279 children. Of those 1,279 children, 653 required placement in departmental custody. The total cost of providing custodial care to children exposed to meth labs in Missouri since August 2005, was approximately $3.4 million according to the department.[23] In one Missouri county, so many children were being removed from meth lab homes and placed in state custody that there are now no longer any foster families available to care for them.[24] Similarly, according to the Tennessee Department of Children's Services, 1,625 children were removed from meth lab homes from January 2007 through December 2011 and placed in foster care at a cost of approximately $70.1 million.[25]

Impacts on the Environment

DEA's Authorized Central Storage Program

In an effort to reduce the cost of lab cleanups and related law enforcement costs, DEA initiated the Authorized Central Storage Program (Container Program) in 2004 to streamline the process. The program allows state and local law enforcement to perform the removal of chemicals from small laboratories, such as One Pots, and temporarily store the chemicals in a centralized location pending final removal by a DEA vendor and final disposal at DEA's expense.

Temporarily storing the waste allows officers to expedite the removal of seized chemicals by eliminating the wait time for vendors to arrive on-site. The program also reduces costs by allowing the vendor to pick up and remove the hazardous waste recovered from multiple sites at centralized locations. According to DEA, this has decreased the average cost per cleanup for small laboratories from $2,300 to less than $500 per lab.

The raw materials and waste of the meth labs pose environmental dangers because they are often disposed of indiscriminately by lab operators to avoid detection, and can also cause residual contamination of exposed surfaces of buildings and vehicles where the meth was being made. According to DEA, for every pound of meth produced, 5 to 6 pounds of toxic waste are produced. Common practices by meth lab operators include dumping this waste into bathtubs, sinks, or toilets, or outside on surrounding grounds or along roads and creeks. Some may place the waste in household or commercial trash or store it on the property. In addition to dumped waste, toxic vapors from the chemicals used and the meth-making process can permeate walls and ceilings of a home or building or the interior of a vehicle, potentially exposing unsuspecting occupants. As a result, the labs potentially end up contaminating the interiors of dwellings and vehicles as well as water sources and soil around the lab site for years if not treated.

Because of the dangerous chemicals used in making meth, cleaning up clandestine methamphetamine labs is a complex and costly undertaking. According to regulations promulgated for the Resource Conservation and Recovery Act by the Environmental Protection Agency, the generator of hazardous waste is the person who produced or first caused the waste to be subject to regulation.[26] The act of seizing a meth lab causes any chemicals to be subject to regulation and thus makes law enforcement the "generator" of the waste when seizing a lab. Accordingly, seizing a lab makes a law enforcement agency responsible for cleaning up the hazardous materials and the costs associated with the cleanup. The materials seized at a clandestine drug laboratory site become waste when law enforcement officials make the determination of what to keep as evidence. Those items not required as evidence are considered hazardous waste and must be disposed of safely and appropriately. The task of removal and disposal of the hazardous waste is usually left to contractors who have specialized training and equipment to remove the waste from the lab site and transport it to an EPA-regulated disposal facility. Depending on the size of the lab, the cost for such a service to respond to an average lab incident can range from $2,500 to $10,000, or up to as much as $150,000 to clean up super labs, according to DOJ.

To help state and local agencies with the expense of lab cleanup, DEA established a lab cleanup program where DEA contracts with vendors and pays them to conduct the cleanup on behalf of the law enforcement agency seizing the lab.[27] In fiscal year 1998, DEA began funding cleanups of clandestine drug labs that were seized by state and local law enforcement agencies, focusing on the removal and disposal of the chemicals, contaminated apparatus, and other equipment.[28] State and local law enforcement agencies seeking to utilize this service contact the DEA to coordinate the cleanup effort. According to DEA program officials, DEA has spent over $142 million on these cleanups nationwide since calendar year 2002. See figure 5.

Given that labs can be placed in a wide range of locations, such as apartments, motel rooms, homes, or even cars, there is also the potential need for further remediation of these areas beyond the initial cleanup of hazardous waste if they are to be safely used or occupied again. Whereas cleanup involves the removal of large-scale contaminants, such as equipment and large quantities of chemicals for the purpose of securing evidence for criminal investigations and reducing imminent hazards such as explosions or fires, remediation involves removing residual contaminants in carpeting or walls, for example, to eliminate the long-term hazards posed by residual chemicals.

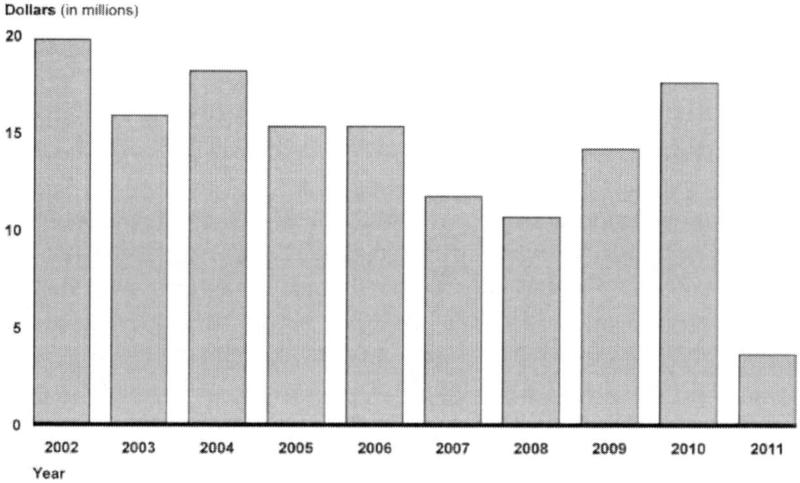

Note: By February 2011, DEA had exhausted all of the $8.3 million in appropriated fiscal year 2011 funds for lab cleanups. The program did not restart operations until October 2011, when fiscal year 2012 began and additional funds became available.

Source: DEA.

Figure 5. DEA Funds Spent on the Lab Cleanup Program, Calendar Years 2002 through 2011.

Procedures for remediation of a property or structure usually involve activities such as the removal of contaminated items that cannot be cleaned, such as carpeting, and wallboard; ventilation; chemical neutralization of residues; washing with appropriate cleaning agents; and encapsulation or sealing of contaminants, among other activities. Depending on the extent of the contamination, the cost to remediate a property can be substantial.[29] Extremely contaminated structures may require demolition. However, unlike the funding that is available for initial lab cleanup from DEA, there are no federal funds available for remediation, leaving the owner of a contaminated property responsible for the costs of any remediation to be done.

Impacts on Law Enforcement

Because of their toxic nature, meth labs pose a serious physical danger to law enforcement officers who come across or respond to them, and therefore must be handled using special protective equipment and training that are costly to law enforcement agencies. The process of cooking meth, which can result in eye and respiratory irritations, explosions and fires, toxic waste products, and contaminated surroundings, can be dangerous not only to the meth cook but also to persons who respond to or come across a lab, such as law enforcement officers. Because of the physical dangers posed by the labs, the Occupational Safety and Health Administration has established requirements for persons, including law enforcement, entering a clandestine lab. 30 These requirements include initial and annual training on hazardous waste operations, annual physical exams to monitor the ongoing medical condition of individuals involved in handling meth lab sites, and guidelines for protective equipment to be used when working in a lab. Consequently, whether the lab is raided by investigators or encountered by accident during the course of an investigation, first responders and police agencies are required to provide their personnel specialized training and equipment, such as hermetically sealed hazmat suits, to safely process a lab.

Because of the complexity involved in handling meth labs, seizing even a small lab can demand significant time and resources of law enforcement agencies. The processing of a lab can take hours and require the involvement of several officers to address the hazards left behind, collect and document evidence, and guard the scene of the lab while it is being processed. For example, according to one law enforcement official in Tennessee, responding to a meth lab requires at least two protectively suited-up officers inside the lab to gather evidence, two additional officers suited up outside the lab as a backup team in case something happens with the lab and they need to respond, and at least one other officer on-site to provide security while the lab is being processed for evidence and cleanup. According to one estimate provided by a law enforcement agency in Indiana, the cost to the agency of the officers' time as well as the protective equipment and processing supplies required to respond to a lab can exceed $2,000 per lab. Given these costs, law enforcement officials from all case study states agreed that responding to meth labs can be a significant financial burden on their agencies. For example, in fiscal year 2010, the Tennessee Meth Task Force spent $3.1 million providing equipment and training to law enforcement personnel and responding to meth lab incidents. Further, unlike large multinational drug-trafficking organizations, meth lab operators are usually lower income and producing meth for personal use; thus operators usually have little in the way of valuable assets or cash that law enforcement agencies can seize as a way of recouping the lab seizure response costs.

ELECTRONIC TRACKING SYSTEMS HELP ENFORCE PSE SALES LIMITS BUT HAVE NOT REDUCED DOMESTIC METH LAB INCIDENTS

Tracking Systems Can Help Enforce Individual Sales Limits, Identify Potential Diversion, and Prosecute Meth Crimes

Electronic tracking systems can help prevent individuals from purchasing more PSE product than allowed by law. By electronically automating and linking logbook information on PSE sales and monitoring sales in real time, stop sale electronic tracking systems can block individuals attempting to purchase more than the daily or monthly PSE limits allowed by federal or state laws. All sales in states using the NPLEx system are linked; thus the system can also be used to block individuals who attempt to purchase more than the allowable amount of PSE in any state using the NPLEx system. According to data provided by the vendor that provides the NPLEx software platform, in 2011, the system was used to block the sale of more than 480,000 boxes and 1,142,000 grams of PSE products in 11 states. Similarly, as of July 31, 2012, the system was used to block the sale of more than 576,000 boxes and 1,412,000 grams of PSE products in the 17 states using the system in 2012. See table 1.

By automating the logbook requirement set forth by the CMEA, electronic tracking systems can make PSE sales information more accessible to law enforcement to help it investigate potential PSE diversion, find meth labs, and prosecute individuals for meth-related crimes. [31] Law enforcement officials we spoke with in all four case study states that use electronic tracking systems reported using the systems for one or more of these purposes. For example, officers from a Tennessee narcotics task force told us how they use the NPLEx system to help identify the diversion of PSE for meth production. According to these officers, the NPLEx system provides them with both real-time and on-demand access to pharmacy logs via a website and includes automated tools that enable them to monitor suspicious buying patterns or specific individuals.[32]

Table 1. PSE Sales and Blocks for NPLEx States, 2011 through 2012

	Number of states on NPLEx	Purchases	Blocksa
2011	11	15,479,147 (15,964,132 boxes; 31,089,533 grams)	435,406 (484,325 boxes; 1,142,032 grams;
January 1, 2012, through July 31, 2012	17	18,729,125 (19,230,688 boxes; 38,741,225 grams)	500,031 (576,840 boxes; 1,412,451 grams)

Source: Appriss.

Note:

[a]Blocks occur when an individual attempting to purchase PSE product has already purchased the maximum amount of PSE product allowed by state or federal law or does not meet other established requirements such as being of minimum age or have been previously convicted of manufacturing methamphetamine, for example. In such instances, the system notifies the sales clerk to stop or "block" the sale from being completed.

In one particular case, the taskforce used NPLEx's monitoring tools to place a watch on a specific individual previously identified as being involved in illegal meth activity. When the individual subsequently purchased PSE, the task force received a notification e-mail of the purchase and upon further investigation was able to determine that the individual had sold the PSE to a Mississippi meth cook. Some law enforcement officials in our four case study states reported that they do not actively use the electronic tracking systems for investigations but rather rely on other sources such as informants, meth hotlines, citizen complaints, and routine traffic stops to identify potential diversion and meth labs. Nevertheless, these officials acknowledged using these systems to obtain evidence needed to prosecute meth-related crimes after meth labs have been found. For example, a law enforcement official in Iowa noted that after officials have identified a suspected lab operator or smurfer, they can use the data in NPLEx to help build their case for prosecution or sentencing by using the records to estimate the amount of PSE that was potentially diverted for meth production. They can also determine for which retailers they need to obtain video evidence to confirm their identity of the individual making the purchase.

Law enforcement officials in Indiana and Tennessee, two states that recently moved from lead-generating systems to the NPLEx stop sale system, reported some challenges with NPLEx as a diversion investigation tool.[33] Prior to the implementation of NPLEx, law enforcement was able to use the lead-generating systems in place to identify individuals who exceeded purchase limits and then take enforcement action or obtain a search warrant based upon the criminal offense. However according to these officials, given that NPLEx blocks individuals from exceeding purchase limits, individuals involved in diversion are no longer as readily identifiable as persons of interest and it now takes longer and is more labor intensive to investigate potential PSE diversions, as they no longer have arrest warrants as a tool to get into a residence suspected of having a meth lab.

Meth Producers Use Smurfing to Circumvent Tracking Systems, and Lab Incidents in Tracking States Have Not Declined

While electronic tracking systems such as NPLEx are designed to prevent individuals from purchasing more PSE than allowed by law, meth cooks have been able to limit the effectiveness of such systems as a means to reduce diversion through the practice of smurfing.[34] Smurfing is a technique meth cooks use to obtain large quantities of PSE by recruiting individuals or groups of individuals to purchase the legal allowable amount of PSE products at multiple stores, and then aggregate for meth production. By spreading out PSE sales among individuals, smurfing circumvents the preventive blocking of stop sale tracking systems.

Meth lab incidents in states that have implemented electronic tracking systems have not declined, in part because of smurfing. For example, meth lab incidents in the three states—Oklahoma, Kentucky, and Tennessee—that have been using electronic tracking systems for the longest period of time are at their highest levels since the implementation of state and federal PSE sales restrictions. While these states experienced initial declines in meth lab incidents immediately following the state and federal PSE sales restrictions put in place from 2004 through 2006, lab incidents have continued to rise since 2007, likely in part because of the emergence of smurfing and the use of the One Pot method for production (see table 2).

Law enforcement officials from every region of the country report that the PSE used for meth production in their areas can be sourced to local and regional smurfing operations.[35] The methods, size, and sophistication of these operations can vary considerably—from meth users recruiting family members or friends to purchase PSE for their own individual labs to larger-scale operations where groups purchase and sell large quantities of PSE to brokers for substantial profits, who in turn often sell the PSE to Mexican drug-trafficking organizations operating super labs in California.[36] Individuals recruited for smurfing have included the elderly, homeless, college students, the mentally handicapped, and inner city gang members, among others.

Table 2. Number of Lab Incidents for States That Have Used Electronic Tracking the Longest, 2002-2011

	2002	**2003**	**2004**	**2005**	**2006**	**2007**	**2008**	**2009**	**2010**	**2011**
Kentucky[a]	388	516	622	616	336	310	**442**	**743**	**1,359**	**1,758**
Oklahoma[b]	1,053	1,426	914	329	**223**	**114**	**194**	**784**	**880**	**1,006**
Tennessee[c]	814	1,589	2,369	1,751	**903**	**603**	**834**	**1,494**	**2,153**	**2,326**

Source: DEA NSS.

Notes:

The number of lab incidents are in bold type for those years that electronic tracking was in effect for the state. Declines in the number of lab incidents that began prior to the states' use of electronic tracking are likely due to legal restrictions on the sale of PSE being put in place through the states and the passage of the CMEA.

[a] Kentucky began using electronic tracking in June 2008.

[b] Oklahoma began using electronic tracking in 2006.

[c] Tennessee began using electronic tracking in 2006.

The use of fake identification by smurfs is an area of growing concern for law enforcement. Smurfs can use several different false IDs to purchase PSE above the legal limit without being detected or blocked by a tracking system. For example, in 2012, through a routine traffic stop, state and local law enforcement officials in Tennessee identified a smurfing ring where a group of at least eight individuals had used more than 70 false IDs over a 9-month period to obtain over 664 grams of PSE. All of the IDs had been used to purchase the maximum amount of PSE allowed, with only one transaction (2.4 grams of PSE) blocked by the electronic tracking system. Law enforcement officials from the four electronic tracking case study states emphasized that investigating smurfing rings can be very time and resource intensive because of the large number of persons involved and the potential use of fraudulent identifications. The use of fake IDs for smurfing can also affect the use of electronic tracking systems as tools to assist in the prosecution of meth-related crimes. According to the National Methamphetamine & Pharmaceuticals

Initiative (NMPI) advisory board, smurfers are increasingly utilizing fake identification and "corrupting" electronic tracking databases to the point where prosecutors prefer eyewitness accounts and investigation (law enforcement surveillance) of violations before filing charges or authorizing arrests or search warrants.[37] This results in costly man-power-intensive investigations.

In summary, based on the experience of states that have implemented electronic tracking, while it has not reduced meth lab incidents overall, this approach has had general impacts, but also potential limitations, including the following:

- Under the current arrangement with CHPA, the operating expenses of NPLEx are paid for by PSE manufacturers and provided to the states at no cost.
- Automating the purchase logbooks required by the CMEA and making the logbook information available in an electronic format to law enforcement is reported to be a significant improvement over paper logs that have to be manually collected and reviewed. This record-keeping ability is reported to have also been useful in developing and prosecuting cases against individuals who have diverted PSE for meth production.
- Electronic tracking maintains the current availability of PSE as an OTC product under limits already in place through the CMEA and related state laws.
- The NPLEx system helps to block attempts by a consumer using a single identification to purchase PSE products in amounts that exceed the legal limit, and can prevent excessive purchases made at one or more locations.
- Although PSE manufacturers currently pay for the NPLEx system, depending on the circumstances, their financial support may not necessarily be sustained in the future.
- Although electronic tracking can be used to block sales of more than the legal amount to an individual using a given identification, through the practice of smurfing, individuals can undermine this feature and PSE sales limits by recruiting others to purchase on their behalf or by fraudulently using another identification to make PSE purchases.
- According to some law enforcement officials, the stop sale approach of the NPLEx system makes it more challenging to use the system as an investigative tool than a lead-generating system because it prevents individuals from exceeding purchase limits, which would otherwise

make them more readily identifiable to law enforcement as persons of interest.

- The practice by smurfers of using fraudulent identification to purchase PSE products has been reported to diminish the ability of electronic tracking systems to assist in the prosecution of meth related crimes. According to some law enforcement officials, the rising use of fraudulent identifications has also increased the need to gather eyewitness accounts or conduct visual surveillance to confirm the identities of the individuals, a development that in turn has been reported to lead to more time- and resource-intensive investigations.

Prescription-Only Appears to Help Reduce Lab Incidents; Full Impact on Consumers Is Unknown and May Be Limited on Health Care System

Implementation of Prescription-Only Approach Followed by Declines in Lab Incidents, although Some PSE Diverted from Other States

The number of reported meth lab incidents in both Oregon and Mississippi declined following the adoption by those states of the prescription-only approach for PSE product sales (see fig. 6). In the case of Oregon, the number of reported meth lab incidents had already declined by nearly 63 percent by 2005 from their 2004 peak of over 600 labs. After the movement of PSE products to behind-the-counter status in Oregon in 2005 and implementation of the CMEA and state-imposed prescription-only approach in 2006, the number of reported meth lab incidents in Oregon continued to decline in subsequent years. In Mississippi, after the adoption of the prescription-only approach in 2010, the number of reported meth lab incidents subsequently declined from their peak by 66 percent to approximately 321 labs in 2011. See figure 6 below.

The communities in Missouri that have adopted local prescription-only requirements also experienced a decline in the number of meth labs. For example, while lab incidents statewide in Missouri increased nearly 7 percent from 2010 to 2011, the area in southeastern Missouri where most of the communities have adopted prescription-only ordinances saw lab incidents decrease by nearly half.

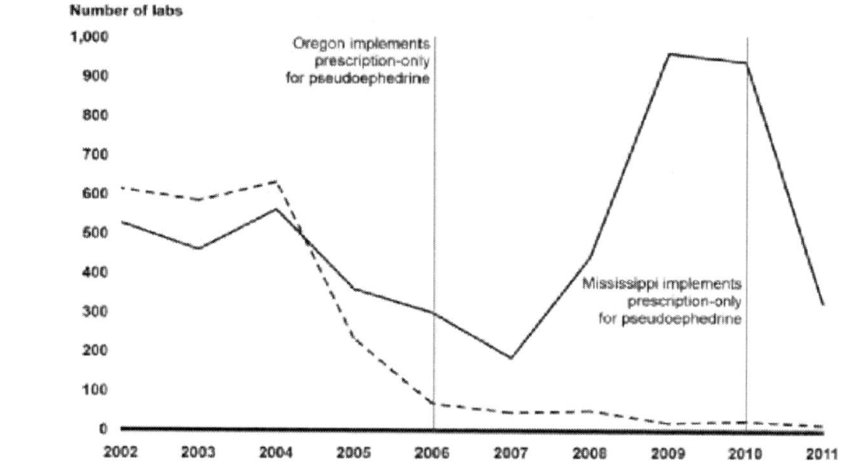

Year	2002	2003	2004	2005	2006	2007	2008	2009	2010	2011
Mississippi	527	459	561	359	299	182	440	960	937	**321**
Oregon	614	584	632	232	67	**43**	**48**	**17**	**21**	**11**

Notes:

The number of lab incidents are in bold type for those years that the prescription-only approach was in effect for the state for a full year. Declines in the number of lab incidents that began prior to the states' use of the prescription-only approach are likely due to legal restrictions on the sale of PSE being put in place through the states and the passage of the CMEA.

Data was accessed on October 1, 2012.

[a]Mississippi implemented the prescription-only approach in July 2010. [b]Oregon implemented the prescription-only approach in July 2006.

Source: GAO analysis of data from DEA's National Seizure System.

Figure 6. Reported Meth Lab Incidents in Oregon and Mississippi, 2002 through 2011.

Even as declines were observed in Oregon and Mississippi after implementing the prescription-only approach, declines were also observed in neighboring states that did not implement the approach, possibly because of other regional or reporting factors. For example, all states bordering Oregon also experienced significant declines in meth labs from 2005 through 2011, ranging from a 76 percent decline for California to a 94 percent decline for Washington state. In Mississippi's case, except for Tennessee, all bordering states also experienced declines in lab incidents from 2009 through 2011, ranging from a 54 percent decrease in Arkansas to a decline of 57 percent in

Louisiana. Consequently, there may be some other factors that contributed to the lab incident declines across all these states regardless of the approach chosen. One potential factor for the declines observed from 2010 through 2011 is the exhaustion of DEA funds to clean up labs. According to DEA officials, as the funds provide an incentive to state and local agencies to report meth lab incidents to DEA, the lack of funds from February 2011 to October 2011 may have resulted in fewer lab incidents being reported during this time period. Other potential factors within the states may have also contributed to declines in the number of lab incidents in neighboring states. For example, Arkansas law enforcement officials reported that in 2011, a change in state law took effect that made it illegal to dispense PSE products without a prescription, unless the person purchasing the product provided a driver's license or identification card issued by the state of Arkansas, or an identification card issued by the United States Department of Defense to active military personnel.[38] In addition, Arkansas law requires that a pharmacist make a professional determination as to whether or not there is a legitimate medical and pharmaceutical need before dispensing a nonexempt PSE product without a valid prescription.[39] As a result of these additional requirements, retailers such as Walmart decided to no longer sell PSE products OTC in Arkansas and instead require a prescription.

According to state and local law enforcement officials in Oregon and Mississippi, the prescription-only approach contributed to the reduction of reported meth lab incidents within those states. For example, according to the executive director of the Oregon Criminal Justice Commission and the directors of the Mississippi Bureau of Narcotics and the Gulf Coast HIDTA, the decline in meth lab incidents in their states can be largely attributed to the implementation of the prescription-only approach. Although their perspectives cannot be generalized across the broader population of local law enforcement agencies, law enforcement officials of other agencies we met with in Oregon and Mississippi also credited the reduction in meth lab incidents to the implementation of the prescription-only approach.

To determine the extent to which the declines in lab incidents in Oregon were due to the prescription-only approach rather than other variables such as regional or reporting factors, we conducted statistical modeling analysis of lab incident data, the results of which indicate a strong association between the prescription-only approach and a decline in meth lab incidents.[40] Specifically, our analysis showed a statistically significant associated decrease in the number of lab incidents in Oregon[41] following introduction of the law, with the lab incident rate falling by over 90 percent after adjusting for other factors.[42]

With the decline in meth lab incidents, officials in the prescription-only states reported observing related declines in the demand and utilization for law enforcement, child welfare, and environmental cleanup services that are needed to respond to meth labs:

- *Law enforcement:* Local law enforcement officials in Oregon and Mississippi reported that the reduction in meth lab incidents has reduced the resource and workload demands for their departments to respond to and investigate meth labs. For example, one chief of a municipal police department in Oregon reported that the decline in meth labs has resulted in reduced costs to his department largely in the form of reduced manpower, training, and equipment expenses and noted that lab seizures are now so rare that his department no longer maintains a specialized team of responders to meth labs. Another chief of a municipal police department in Mississippi noted that since the adoption of the prescription-only approach, the amount of time and resources spent on meth-related investigations has declined by at least 10 percent.
- *Child welfare*: Officials in both Oregon and Mississippi reported a reduction in the demand for child welfare services to assist children found in households where meth lab incidents occurred. For example, according to a coordinator in Oregon's Department of Human Services, the state has not removed a child from a household with an active lab since 2007. In Mississippi, the Methamphetamine Field Coordinator with the state Bureau of Narcotics, which tracks the number of drug-endangered children for the state, reported that the number of such children declined by 81 percent in the first year that the prescription-only approach was in effect.
- *Environmental cleanup*: According to data from DEA and the Oregon Department of Environmental Quality, declines in costs to clean up labs in Oregon occurred prior to the implementation of the prescription-only approach, falling from almost $980,000 in 2002 to about $580,000 in 2005. However, since 2006, costs for lab cleanup continued to fall and were about $43,000 in 2011. Funding for cleanups in Mississippi showed more variation and fluctuation from year to year; however, between 2010, when the prescription-only

approach was implemented, and 2011, cleanup costs dropped by more than half (from over $1 million to less than $400,000).

However, even as the prescription-only approach appears to have contributed to reducing the number of lab incidents in Oregon, the availability and trafficking of meth is still widespread and a serious threat in the state. According to a threat assessment by the Oregon HIDTA, while the number of reported meth lab incidents has declined, crystal meth continues to be highly available in the area as Mexican drug traffickers import the finished product from laboratories outside the state and from Mexico.

Moreover, while the prescription-only approach appears to have contributed to a reduction in the number of meth labs in the states that have adopted it, the experience of these states to date has shown that the prescription-only approach does not preclude individuals from traveling to neighboring states to purchase PSE products for use in meth labs. Consequently, even as the number of meth lab incidents has declined in prescription-only states, law enforcement reports that many lab incidents that still occur in these states are largely due to PSE product obtained from states without a prescription requirement for PSE. For example, according to a threat assessment by NDIC, law enforcement officers interviewed in 2011 reported that the more stringent restrictions on pseudoephedrine sales in Mississippi have led many pseudoephedrine smurfing groups to target pharmacies in the neighboring states of Alabama, Louisiana, and Tennessee in order to continue operations. Officials of a sheriff's office in a county located along the Gulf Coast in Mississippi stated that the department's investigations have found that large numbers of individuals from Mississippi travel out of state to purchase PSE in an effort to circumvent the Mississippi prescription-only law. While some out-of-state purchases may be for licit uses, the officials stated that they believed a substantial proportion of the PSE brought back from other states was likely being diverted for the production of meth. According to law enforcement officials in Oregon, most of the incidents reported there in recent years involved either dumpsites or inactive "boxed labs" that had been used in previous years but have been dismantled and stored away for potential future use. According to the legal counsel for the Oregon Narcotics Enforcement Association, the association asked law enforcement to determine the source of PSE for lab incidents, in cases where that could be determined. In every case where a determination could be made, it was reported that the PSE was obtained from neighboring states, mostly Washington, but also Idaho, California, and Nevada.

According to PSE purchase activity data from the NPLEx electronic tracking system and the vendor that provides its software platform, individuals using Oregon identifications have purchased PSE products in neighboring states.[43] These data indicate that from October 15, 2011, through August 31, 2012, over 30,000 purchases were made by individuals using Oregon identifications. Similarly for Mississippi, reports by law enforcement of individuals traveling to neighboring nonprescription-only states to purchase PSE products is supported by PSE purchase activity data provided by the NPLEx electronic tracking system.[44] Since the time the NPLEx system has been implemented in these states, the PSE purchase activity data indicate that over 172,000 purchases have been made by individuals using Mississippi identifications.[45]

Some states have taken action to eliminate the opportunity for out-of-state individuals to purchase PSE products without a prescription. For example, in 2011, Arkansas passed legislation making it illegal to dispense any PSE product without a prescription, unless the purchaser can provide an Arkansas-issued driver's license or identification card, or an identity card issued by the U.S. Department of Defense for active-duty military personnel.[46] In 2012, Alabama enacted a similar law that requires individuals residing within a state that requires a prescription to obtain a PSE product to provide a valid prescription for the PSE products if they seek to obtain the products in Alabama.[47] In essence, the impact of these laws is to extend the prescription-only requirement for Mississippi residents into Arkansas and Alabama. Officials from the Mississippi Bureau of Narcotics said these laws will help prevent PSE product from being obtained and diverted to Mississippi for use in meth labs.

In addition to obtaining PSE products from non-prescription states, another potential source of PSE for meth labs in prescription-only states and localities is through the illicit diversion of PSE obtained with a prescription. Similar to techniques used to divert other controlled prescription drugs such as pain relievers, diversion of prescribed PSE can occur through prescription forgery, illegal or improper prescribing by a physician, or "doctor shopping," where an individual goes to several doctors to obtain a prescription from each doctor. Although these may provide potential sources of PSE for use in meth labs in prescription-only states, law enforcement officials in Oregon and Mississippi reported no known instances from their meth lab investigations of finding that a PSE product was obtained through one of those methods in order to make meth. Law enforcement officials in Missouri localities where the prescription-only requirement has been adopted reported a few instances of

PSE obtained with a prescription being used to make meth. According to investigators from a regional drug task force in a county in Missouri, they have found PSE obtained by prescription in at least three meth lab incidents. Since the county has adopted the prescription-only approach, they are observing more instances in which prescriptions of PSE are found at lab incidents. However, they did not find any evidence in these cases that the PSE had been prescribed illegally or obtained through prescription forgery or doctor shopping.

Prescription-Only Status for PSE Appears to Have Reduced Sales of PSE Products, but the Full Impact on Consumers Is Not Known

Judging from the experience of Mississippi, the volume of PSE products obtained by consumers after the adoption of the approach declined from levels that existed when PSE was available OTC. Data on Mississippi OTC PSE product sales and the number of prescriptions for PSE filled suggest that use of PSE products could have fallen by several hundred thousand units after the implementation of the prescription-only approach. For example, annual unit sales of PSE dropped from almost 749,000 in 2009 before the prescription-only approach went into effect, to approximately 480,000 total units of PSE product sold OTC or prescribed in 2010, when the approach was in effect for half the year, to approximately 191,000 units prescribed or sold during 2011, when the approach had been in place for the full year (see table 3).[48] Data are not available for Oregon on the sales of PSE products immediately before and after the implementation of the prescription-only approach to do a comparable analysis.[49]

Given the more restrictive access to PSE products consumers would face under the prescription-only approach, it is expected that consumers will be impacted. The extent of this impact depends on a number of variables such as the potential change in the effective price of PSE that the requirements of the prescription-only approach result in and the availability of effective substitutes or alternative remedies for PSE, for example. Under the prescription-only approach, the effective price of PSE, which includes costs associated with obtaining a prescription, such as the costs of time and travel to the physician for an appointment as well as any associated copays or out-of-pocket charges for the appointment itself, would increase if an in-person visit were necessary, having a negative impact on consumers.

Table 3. Units of PSE Products Sold OTC or Prescribed in Mississippi, January 2009-December 2011

	52 weeks ending 12/27/2009		**52 weeks ending 12/26/2010**		**52 weeks ending 12/25/2011**	
	Dollar sales	**Units sold or prescribed**	**Dollar Sales**	**Units sold or prescribed**	**Dollar sales**	**Units sold or prescribed**
OTC PSE sales	$7,374,772	731,522[a]	$4,011,158	383,313[a]	$21,341	a, b 3,328
Percentage change from previous period	Not applicable.	Not applicable.	-45.6	-47.6	-99.5	-99.1
Percentage change from 12/27/2009	Not applicable.	Not applicable.	-45.6	-47.6	-99.7	-99.5
Prescriptions filled for PSE products	Not available.	17,221 a, c	Not available.	96,454[a]	Not available.	187,680[a]
Total PSE prescriptions filled and OTC sales	Not available.	748,743	Not available.	479,767	Not available.	191,008

Source: GAO analysis of data from IMS Health Incorporated and the Mississippi Prescription Drug Monitoring Program.

[a]Values shown are for the number of units of OTC PSE product sold or prescriptions filled for drugs containing PSE. Although the amount of PSE sold OTC or dispensed by prescription in any given unit can vary depending on the size of the OTC package or prescription, for the purposes of this analysis all units reported sold OTC or prescribed are assumed to contain equivalent quantities of PSE.

[b]According to officials that provided these data, OTC PSE product sales may have been reported after the prescription-only approach was in effect because of a number of potential reasons: sampling error, data entry error, clerks overriding the system at the point of sale when a prescribed PSE purchase was being made but the system did not recognize it, or the possibility of retailers making an OTC sale of a PSE product despite the prescription-only requirement.

[c]Prescriptions filled for PSE recorded before the implementation of the prescription-only approach in July 2010 reflect prescriptions for medications for which PSE is included as a combination ingredient with another ingredient for which a prescription is needed.

If the PSE prescriptions are being obtained by consumers at a higher effective price because of these factors, consumers can be expected to be negatively impacted to some extent by the prescription-only approach. At the same time, some of these costs, such as the costs of time and travel to go to an in-person appointment can be mitigated to the extent patients can obtain a prescription for PSE through a telephone consultation with their physicians.

While it is likely that the effective price for PSE products is higher under the prescription-only approach, data on the cost to consumers for obtaining these prescriptions are not available to make this comparison. Further complicating the determination of the change in the effective price of PSE is the fact that the actual costs to a given consumer for that person's time, travel, and insurance coverage can vary from consumer to consumer depending on the person's individual circumstances. For example, given their lack of insurance, uninsured consumers or patients will likely face higher effective costs to obtain PSE products under a prescription-only approach than those with insurance. Because of the uncertainty involving these variables and factors, it is not possible to determine the magnitude of the change in effective price of PSE for consumers.

Despite the likely increase in the effective price of PSE because of the prescription-only approach, according to state agencies and consumer groups, consumers in Oregon and Mississippi have made few complaints about the approach since its implementation, although research or surveys on the issue have not been conducted. For example, according to the executive director of the Oregon Board of Pharmacy, the state agency that adopted the rule making PSE a controlled substance, the board initially received a small number of complaints from consumers when PSE was initially scheduled, but after a number of months, the board stopped hearing about it. Officials at the Mississippi Board of Pharmacy also noted that they have not received any complaints from consumers about the prescription requirement since it went into effect. According to consumer and patient advocacy organizations such as the National Consumers League and the Asthma and Allergy Foundation of America, which conducted surveys of consumers regarding access to PSE products in 2005 and 2010 respectively, neither organization has received feedback or complaints from consumers or patients from either state about the diminished access imposed on PSE products by the prescription-only approach. Both organizations also noted that they have not conducted any additional research or surveys on the issue since their earlier surveys in 2005 and 2010.

Another variable that determines the impact of the prescription-only approach on consumers is the availability of substitutes for PSE that consumers can use as alternatives to offset any potential increase in the effective price to consumers for obtaining PSE by prescription. To ensure that consumers still have access to an unrestricted oral OTC decongestant, manufacturers of cold and allergy medicines reformulated many products by substituting the ingredient phenylephrine (PE), an alternative oral decongestant also approved by FDA for use in OTC medicines that cannot be used to make methamphetamine. However, according to sales data on PE products in Mississippi for the periods before and after implementation of the prescription-only approach, the changes in sales volume for PE products do not appear to show any direct substitution of PE for PSE by consumers. In fact, the change in volume in PE products shows a decrease for the 52-week period ending in December 2011 (see table 4).

Table 4. Sales of OTC PE Products in Mississippi, January 2009-December 2011

	52 weeks ending 12/27/2009		52 weeks ending 12/26/2010		52 weeks ending 12/25/2011	
	Dollar sales	Units sold	Dollar sales	Units sold	Dollar sales	Units sold
PE sales	$6,280,693	1,137,699[a]	$6,323,837	1,155,900[a]	$6,042,349	1,087,800[a]
Percentage change from previous period	Not applicable.	Not applicable.	0.7	1.6	-4.5	-5.9
Percentage change from 12/27/2009	Not applicable.	Not applicable.	0.7	1.6	-3.8	-4.4

Source: GAO analysis of IMS Health Incorporated data.

[a] Although the amount of PE sold in any given unit can vary depending on the size of the package, for the purposes of this analysis all units reported sold are assumed to contain equivalent quantities of PE.

The lack of a consumer shift from PSE products to PE products could be the result of several potential factors, but data are limited or unavailable to ascertain their impact. For example, it could reflect, on average, consumer perception that PE is not an effective substitute for PSE. Similarly, it could also be an indication that consumers are choosing to forgo medicating their conditions or are using other medications or remedies to relieve their symptoms. Another potential factor that could contribute to this lack of a consumer shift to PE from PSE is the extent to which PSE sales were being diverted for meth use. Although available estimates of the extent to which PSE sales are being diverted vary greatly, the drop in PSE sales without a corresponding increase in PE product sales could also imply that some of the

PSE sales were likely being diverted for meth production. According to officials of the market research firm that provided the PE sales data, another potential explanation for the lack of a distinct shift in demand for PE is the fact that several PE products had to be recalled by the manufacturer because of manufacturing issues.

Industry has noted that PE has limitations as a direct substitute for PSE, and in 2007, FDA reexamined the effectiveness of PE at the approved dosing levels. At the request of citizen petitioners who claimed that the available scientific evidence did not demonstrate the effectiveness of PE at the approved 10-miligram dosage level, an FDA advisory committee reviewed the issue in December 2007, including two meta-analyses of studies provided by the citizen petitioners and CHPA.[50] After reviewing this evidence, the committee concluded that, while additional studies would be useful to evaluate higher doses, the 10-miligram PE dose was effective. However, since the recommendation of the FDA advisory committee in 2007 to study the effectiveness of PE at higher dosage levels, it appears that limited work has been undertaken to do so.[51] According to CHPA, while it agrees that the approved dosage levels of PE are effective, PE has known limitations that make it a less than viable substitute for PSE in some long-duration applications and for many consumers.[52]

As would be expected under the more restrictive prescription-only approach, consumers of PSE products would be negatively impacted to some extent by enactment of the prescription-only approach, considering the variables that determine the change in the effective price of PSE products. However, because of uncertainties related to these variables, such as consumers' individual situations regarding insurance or the need for an in-person consultation with their physicians, the effectiveness of substitutes such as PE or use of other alternatives, and the extent to which PSE sales have been used for illicit purposes, the net effect on consumer welfare resulting from enactment of a prescription-only policy cannot be quantified.

The Impact on the Health Care System Is Generally Unknown but Substantial Changes in Workload Have Not Been Reported to Date

One of the concerns expressed by industry about the potential impact of the prescription-only approach is that it is likely to increase the workload of health care providers and the overall health care system to some extent. Both the Oregon and Mississippi laws required individuals to obtain a prescription from a health care provider which requires some type of visit or consultation with the provider. This visit or consultation requires increased provider workload to process the prescriptions. In addition, individuals who do not already have an established relationship with a health care provider may require a more involved, initial in-person visit to obtain a prescription, and pharmacies may experience increased workload because of new dispensing requirements. Assuming that health care providers charge prices that reflect the costs of providing these additional services, any increase in the workload of health care providers should get reflected in the office charge billed the patient.

While the impact of the prescription-only approach on the health care system is generally unknown, on the basis of limited information available from health care providers in Oregon and Mississippi, it does not appear that there has been a substantial increase in workload demands to provide and dispense prescriptions for PSE products. According to a 2011 study commissioned by CHPA on managing access to PSE, judging from Oregon's experience, the number of health care provider visits did not grow significantly, as consumers have noted obtaining a prescription via telephone or fax request. Officials from associations representing physicians in Oregon stated that their members have not reported any real impact on their practices, and their research from members suggests that the benefits of fewer meth labs outweighs any inconvenience for their membership of requests for prescriptions. Officials from the association representing Mississippi physicians similarly reported that from the perspective of a limited sample of its members involved in family practice, emergency room care, and addiction treatment, no increase had been observed in the demand for appointments from patients seeking PSE products.

In addition, representatives from the association representing pharmacists in Oregon stated that they have received few complaints over the prescription-only requirement. Further, reports from the experience of Oregon and Mississippi indicate that there has not been a significant increase in cost to the states' Medicaid programs. In terms of an impact on states' Medicaid programs, officials in those states said there was no net change in their programs' policy with the implementation of the prescription requirement statewide because their programs already required that participants obtain a prescription for PSE products if they wanted to have the medication covered under the states' Medicaid pharmacy benefit formulary.[53]

In summary, on the basis of the experience of Oregon and Mississippi, the use of the prescription-only approach has had the following impacts:

- Its apparent effectiveness in reducing the availability of PSE for meth production has in turn helped to reduce or maintain a decline in the number of meth lab incidents in the states that have adopted the approach.
- The reduction in meth lab incidents has led to a corresponding decline in the demand or need by communities for child welfare, law enforcement, and environmental cleanup services to respond to the secondary impacts of the meth labs.
- Although it is difficult to quantify due to the lack of data and wide variation in the individual circumstances of consumers, the prescription-only approach has the potential for placing additional burdens on consumers to some extent.
- Increased the potential for additional workload and costs for the health care system to provide prescriptions for PSE products. From the limited information and data that are available to date, it is not clear that they have been substantial in the two states that have adopted the prescription-only approach to date.
- Increased possibility of consumers in a prescription-only state attempting to bypass the prescription-only requirement by purchasing PSE in a neighboring nonprescription state.

AGENCY COMMENTS

We provided a draft of this report to the Department of Justice and ONDCP for comment. Justice and ONDCP did not provide written comments on the report draft, but both provided technical comments, which we incorporated as appropriate.

Carol R. Cha,
Acting Director
Homeland Security and Justice

APPENDIX I: OBJECTIVES, SCOPE, AND METHODOLOGY

Our objectives were to identify (1) trends in domestic meth lab incidents over the last decade and the impact of domestic meth labs on the communities affected by them; (2) the impact of electronic tracking systems on domestic meth lab incidents and the limitations, if any, of using these systems; and (3) the impact of prescription-only laws on domestic meth lab incidents and any implications of this approach for consumers and the health care system.

To identify the trends in domestic meth lab incidents over the last decade, we obtained and analyzed data for all states from the Drug Enforcement Administration's (DEA) National Seizure System (NSS) data on nearly 149,000 lab seizure incidents that occurred during the last 10 calendar years, 2002 through 2011.[1] Using these data, we analyzed the number of incidents nationally and by region and by type of lab (i.e., P-2-P, Nazi/Birch and One Pot, or Red Phosphorus) and lab capacity. To assess the reliability of these data, we discussed the sources of the data with agency officials knowledgeable about the data to determine data consistency and reasonableness and compared them with other supporting data and documentation, where available, from states selected to be case studies for this review. As reporting by state and local law enforcement agencies of lab incidents to DEA is voluntary except when DEA provides funds to the agencies for lab cleanup, because of the exhaustion of DEA's cleanup funds less than half way through fiscal year 2011, the number of lab incidents reported for 2011 could be biased downward as compared with the number of incidents in previous years. We discussed this issue and its potential implications with DEA officials that manage the collection of the data and the steps they have taken to address it. From these efforts and discussions, we determined that the data were sufficiently reliable for the purposes of this report.

To identify key factors that influenced lab seizure incident trends over time, we obtained the perspectives and information on meth lab incident trends and factors influencing this trend from state and local officials we interviewed in states that were selected as case studies. This nonprobability sample of states was selected to reflect a mix of characteristics such as the type of approach chosen for controlling the sale of pseudoephedrine (PSE) products (electronic tracking or prescription-only), length of time the approach has been in use, and the number of meth labs seized relative to the state's population size. The states selected for inclusion as case study states included Iowa (electronic tracking), Kentucky (electronic tracking), Mississippi (prescription-only), Missouri (electronic tracking), Oregon (prescription-only), and

Tennessee (electronic tracking). While we cannot generalize any findings or results to the national level from our sample of states visited for our case studies, the information from these states provided perspective on meth lab trends and the experiences of the states in implementing these approaches. We also reviewed drug threat assessments and reports by the National Drug Intelligence Center (NDIC) and information from officials with DEA and the Office of National Drug Control Policy (ONDCP). We reviewed the methodology of the assessments and reports and found them sufficiently reliable to provide perspectives on meth lab incident trends and factors influencing these trends. We obtained additional information and input regarding factors that contributed to meth lab incident trends from federal, state, and local officials participating in the May 2012 conference of the National Methamphetamine and Pharmaceutical Initiative (NMPI), a national initiative funded by ONDCP.

To determine the impact of domestic meth labs on the communities affected by them, we first reviewed a variety of reports and studies on meth labs and their impacts from sources such as the Department of Justice (DOJ), DEA, the RAND corporation, media reports, and published academic research to identify the particular areas or ways that communities are directly affected as a result of the presence of labs. On the basis of this review, we identified the key ways communities are impacted by meth labs. These included the provision of health care to meth lab burn victims, threats and dangers posed to the welfare of children, environmental damage, and increased demand and workload for law enforcement agencies. While there are other areas or ways that can be impacted by meth labs, such as treatment for health-related conditions related to meth abuse and the demand for addiction treatment, these impacts are caused by the abuse of both imported and domestically produced meth and are not impacts unique to meth labs. Therefore, we did not include those areas in our review. To describe impacts on health care providers to administer care to meth lab operators injured or burned by their labs, we reviewed and synthesized information from published academic research comparing the injuries and treatment provided to meth-lab-burn victims as compared with non-meth lab burn related patients, documentation from DOJ on meth labs, and media reports on the reported impacts of meth labs on hospital burn centers. We also interviewed the director of the burn center at the Vanderbilt University Hospital in Nashville, Tennessee, to get his perspective, as the center has treated a significant number of burn patients that received their injuries from a meth lab. To describe impacts of meth labs on child welfare, we reviewed and synthesized information from DOJ on drug-endangered children, meth lab incident data from DEA on the number of children reported to be affected by the labs, and published academic research on the impact of meth abuse on the need for foster care. To describe environmental damage caused by meth labs, we reviewed and synthesized information from DOJ on the impact of meth labs, DEA's guidance for meth lab cleanup, and a report from the DOJ Inspector General on DEA's meth lab cleanup program. For context, we also obtained information from DEA on its clandestine lab cleanup program and the funds expended on the program to assist state and local law enforcement agencies in cleaning up meth labs from 2002 through 2011. In addition, we obtained and analyzed information from the case study states of Mississippi, Missouri, and Oregon on any funds state agencies spent on the cleanup of meth labs. To describe impacts of meth labs on law enforcement agencies in communities, we reviewed and synthesized information from DEA's guidance for meth lab cleanup, documentation from DOJ on meth labs, as well as information from state and local law enforcement officials we interviewed from our case study states.

To determine the impact of electronic tracking systems on domestic meth lab incidents, we analyzed DEA NSS data on the number of meth lab incidents that were reported in the 3 states that have implemented electronic tracking the longest—Kentucky, Oklahoma, and Tennessee— from 2002 through 2011 to identify any trends in lab incidents that occurred within those states before and after the implementation of electronic tracking within those states. To examine the volume of PSE sales activities the national electronic tracking system monitors and blocks when necessary, we obtained and reviewed PSE purchase activity data (purchases, blocks, and exceedances) for 2011 and 2012 from Appriss, the software firm that developed and manages the software program MethCheck, which is used as the operational platform for the National Precursor Log Exchange (NPLEx), the interstate electronic tracking system paid for by manufacturers of PSE products. We chose this time period because those were the most recent years for which data from multiple states were available. To assess the reliability of these data, we discussed the data with Appriss officials. From these efforts and discussions, we determined that the data were sufficiently reliable for the purposes of this report. To understand how electronic tracking works in practice and the limitations of this approach, we obtained information from officials with Appriss as well as officials with state and local law enforcement and the High Intensity Drug Trafficking Areas (HIDTA) in our electronic tracking case study states of Iowa, Kentucky, Missouri, and Tennessee. For these state and local law enforcement officials, we utilized a snowball sampling methodology in which we initially contacted key law enforcement officials in those states involved in dealing with the meth lab problem who identified and provided contacts for other officials in those states to meet with. From these state and local law enforcement officials, we obtained information and their perspectives on the use of electronic tracking, its impact on the meth lab problem within their jurisdictions, and any potential advantages or limitations of the approach identified through their investigations and experience with the system to date. Although their perspectives cannot be generalized across the broader population of state and local law enforcement agencies in electronic tracking states, their perspectives provided insights into and information on the use and impact of the approach in practice and its limitations.

To determine the impact of prescription-only laws on domestic meth lab incidents and any implications of adopting this approach for consumers and the health care system, we analyzed DEA NSS data on the number of meth lab incidents that were reported in the prescription-only states of Mississippi and Oregon and their border states (Alabama, Arkansas, California, Idaho, Louisiana, Nevada, Tennessee, and Washington) from 2002 through 2011 to identify any trends in lab incidents that occurred within those states before and after the implementation of the prescription-only approach. To determine the impact of the prescription-only approach on meth lab incidents in Oregon, we conducted a statistical modeling analysis of the lab incident data that controlled for other factors such as region of the country, ethnic composition of the state population, the proportion of the state population that is male, distance from the Mexican border, and the state drug arrest rate, among others. For more details on the methodology used for this analysis, see appendix III.

To determine the impact of the prescription-only approach in counties and localities in Missouri that have adopted the approach, we also obtained and analyzed information from local officials in Missouri on how meth lab incidents have been impacted since the adoption of the approach within their jurisdictions. To obtain the perspective of state and local officials on the impact of the implementation of the prescription-only approach in their states and localities, we utilized a snowball sampling methodology in which we initially contacted key law enforcement officials involved in dealing with the meth lab problem or associations representing law enforcement in Mississippi, Missouri, and Oregon who then identified and provided contacts of other officials within their states for us to meet with. We interviewed these officials to obtain their perspectives on the impact of the prescription-only approach on the meth lab problem as well as the perceived impacts on other areas, where possible, such as the demand for law enforcement, child welfare, environmental cleanup, and the trafficking of meth within their states. Although their perspectives on these impacts cannot be generalized across the broader population of state and local law enforcement agencies in prescription-only states, their perspectives provided insights into and information on the impact of the approach in practice. To determine the extent to which individuals in prescription-only states have been traveling to neighboring states to obtain PSE product without a prescription or have diverted PSE product obtained with a prescription, we interviewed and obtained information from local law enforcement officials in Mississippi, Missouri, and Oregon on what they have found in their investigations into meth labs and PSE smurfing. We also obtained and reviewed NPLEx data on PSE purchase activity from Appriss for PSE purchases made in Washington state with identifications issued by Oregon from October 15, 2011, to the most recent full month available (August 2012). We chose the starting date of October 15, 2011, because that was the date that Washington state implemented the NPLEx system statewide.

To gauge the extent of PSE sales made in other states neighboring Oregon (California, Idaho, and Nevada) to individuals using identifications issued by Oregon that had not implemented NPLEx but had retailers within the states that used the NPLEx MethCheck software program, we obtained and reviewed the MethCheck log data on PSE purchase activity for those states for the same October 15, 2011, to August 2012 time period. For Mississippi, we obtained and reviewed NPLEx data on PSE purchase activity for purchases made with identifications from Mississippi in the NPLEx states neighboring Mississippi (Alabama, Louisiana, and Tennessee) between the time those states joined NPLEx to July 2012.[2] To determine the impact of the prescription-only approach on consumers in Mississippi, we obtained data from IMS Health Inc. through DEA on the volume of PSE sales for three 52-week periods ending in December 2009, 2010, and 2011 and analyzed the data for any changes in volume over time, comparing the 2010 and 2011 periods when the prescription-only approach was in effect with the 2009 period when it was not. To assess the reliability of the data, we reviewed documentation and information from IMS Health officials knowledgeable about the data to determine data consistency and reasonableness. From these reviews, we determined that the data were sufficiently reliable for the purposes of this report. Because data prior to the period Oregon implemented the prescription-only approach in 2006 were not available, we were not able to do a similar analysis for Oregon.

To examine the number of prescriptions filled in Mississippi for PSE medications, we obtained and reviewed data provided by the Mississippi Board of Pharmacy's Prescription Drug Monitoring Program. To assess the reliability of the data, we discussed the data with officials that manage the program. From these efforts and discussions, we determined that the data were sufficiently reliable for the purposes of this report. To obtain additional information on the reported and estimated impacts of the prescription-only approach on consumers and the health care system, we reviewed a report of the potential impacts of the prescription-only approach prepared for the Consumer Healthcare Products Association (CHPA). To help obtain perspective on the potential impact on consumers, we asked the state boards of pharmacy and state associations representing pharmacists in Mississippi and Oregon, such as the Oregon State Pharmacists Association, about the extent to which complaints may have been made by consumers about the prescription-only approach. We also asked the National Consumers League and the Asthma and Allergy Foundation of America if they had received feedback or complaints from consumers on the impact of the prescription-only approach. We chose these organizations because they have previously surveyed consumers about access to PSE products.

To understand the prescription-only approach's impact on the workload demands for physicians, we obtained the perspective of state associations representing physicians practicing in Oregon and Mississippi, such as the Oregon Medical Association and the Mississippi State Medical Association, on the extent to which their members have reported seeing an increase in demand for appointments for PSE prescriptions and any corresponding increase in their workload. While their perspectives cannot be generalized to the larger population of physicians in these states, they provided insights into the impact of the approach on their members' practices. To determine the impact of the prescription-only approach on the Medicaid programs within Mississippi and Oregon, we obtained perspectives and information from Medicaid program officials in those states on what, if any, changes the approach required of their prescription formulary and any resulting changes in program costs.

We conducted this performance audit from February 2012 to January 2013 in accordance with generally accepted government auditing standards. Those standards require that we plan and perform the audit to obtain sufficient, appropriate evidence to provide a reasonable basis for our findings and conclusions based on our audit objectives. We believe that the evidence obtained provides a reasonable basis for our findings and conclusions based on our audit objectives.

APPENDIX II: STATE LEVEL DATA ON METH LAB INCIDENTS, 2002 THROUGH 2011

State	2002	2003	2004	2005	2006	2007	2008	2009	2010	2011	Total
Alabama	343	524	803	529	273	249	624	673	719	293	5,030
Alaska	35	54	120	66	20	7	18	13	22	5	360
Arizona	294	260	221	138	48	23	34	24	18	5	1,065
Arkansas	646	1,171	1,361	701	450	380	418	671	824	308	6,930
California	1,764	1,303	855	510	444	307	398	301	206	122	6,210
Colorado	527	523	419	274	136	72	62	48	32	13	2,106
Connecticut	2	1	0	4	3	0	1	2	1	0	14
Delaware	0	2	3	1	0	0	1	1	3	2	13
District of Columbia	0	0	1	0	0	0	0	0	1	1	3
Florida	188	319	438	470	202	186	214	415	526	161	3,119
Georgia	224	440	547	433	191	118	197	217	332	141	2,840
Hawaii	12	5	17	18	4	1	0	0	3	0	60
Idaho	134	121	75	35	23	23	14	17	19	8	469
Illinois	711	1,084	1,582	1,430	863	399	379	416	476	637	7,977
Indiana	753	1,046	1,384	1,506	838	815	739	1,328	1,243	1,437	11,089
Iowa	925	1,473	1,687	914	364	198	241	336	380	413	6,931
Kansas	792	705	650	417	194	101	161	184	241	202	3,647
Kentucky	388	516	622	616	336	310	442	743	1,359	1,758	7,090
Louisiana	146	136	178	144	28	54	45	163	218	70	1,182
Maine	0	0	4	5	5	1	4	1	4	5	29
Maryland	1	2	3	5	7	2	2	0	3	1	26
Massachusetts	2	2	2	8	3	4	3	3	2	2	31
Michigan	264	376	461	511	290	212	456	716	866	438	4,590
Minnesota	336	483	288	169	69	48	46	31	27	9	1,506
Mississippi	527	459	561	359	299	182	440	960	937	321	5,045
Missouri	2,765	2,897	2,924	2,340	1,326	1,292	1,520	1,810	1,979	2,114	20,967
Montana	104	131	107	36	13	10	10	18	21	11	461
Nebraska	373	294	327	287	35	30	67	40	27	19	1,499
Nevada	106	249	153	86	44	24	17	16	13	16	724
New Hampshire	1	2	2	9	6	3	1	7	8	15	54
New Jersey	3	1	3	4	6	1	4	0	1	0	23
New Mexico	170	305	226	103	52	45	73	67	64	21	1,126
New York	30	35	70	26	42	13	18	17	33	46	330
North Carolina	73	223	473	493	216	161	197	213	237	400	2,686
North Dakota	211	260	238	175	43	27	35	35	8	9	1,041
Ohio	141	227	535	669	375	232	260	344	381	364	3,528

State	2002	2003	2004	2005	2006	2007	2008	2009	2010	2011	Total
Oklahoma	1,053	1,426	914	329	223	114	194	784	880	1,006	6,923
Oregon	614	584	632	232	67	43	48	17	21	11	2,269
Pennsylvania	31	65	138	101	65	18	24	44	39	9	534
Rhode Island	4	1	0	0	2	0	0	0	0	1	8
South Carolina	70	169	343	253	112	68	130	244	344	338	2,071
South Dakota	38	49	36	26	15	13	11	9	22	5	224
Tennessee	814	1,589	2,369	1,751	903	603	834	1,494	2,153	2,326	14,836
Texas	682	868	740	442	188	158	250	273	192	88	3,881
Utah	153	113	105	67	39	8	15	11	10	9	530
Vermont	0	0	1	2	0	2	0	0	4	0	9
Virginia	10	46	110	87	22	25	21	29	106	202	658
Washington	1,441	1,008	962	547	337	240	127	70	46	33	4,811
West Virginia	67	106	328	445	166	113	116	139	207	92	1,779
Wisconsin	95	128	110	80	33	7	16	27	45	41	582
Wyoming	68	36	27	13	6	9	6	0	11	2	178
Total	18,131	21,817	24,155	17,866	9,426	6,951	8,933	12,971	15,314	13,530	149,094

Source: DEA NSS.

Note: According to DEA officials, records of incidents can be updated within the NSS or new records added as new data and information becomes available or is submitted. Consequently, the number of total number of incidents may vary over time. The data presented above were pulled from the NSS on October 1, 2012.

APPENDIX III: STATISTICAL MODEL ANALYSIS OF THE PRESCRIPTION-ONLY REQUIREMENT ON METH-LAB INCIDENTS

Evaluation of the Impact of Prescription-Only Requirement

We evaluated the impact of the prescription-only pseudoephedrine requirement on domestic production of methamphetamine separately using state-level data. We chose Oregon which implemented its prescription-only pseudoephedrine requirement in 2006. In order to evaluate the impact of the policy, we performed multivariate regression analyses using generalized estimating equations (GEE) to compare the trend in lab seizures, reported to DEA between 2002 and 2010. We compared the case study state with a selected group of control states using a method that improves upon the commonly used Difference-inDifferences (DD) estimation method. [1] We estimated robust standard errors for the DD coefficients by modeling the covariance structure in the GEEs. In addition to estimating a DD model, we, alternatively, estimated the intervention effect by comparing the case state to a single synthetic control using the synthetic control methods in comparison case studies following Abadie and colleagues[2] and Nonnemaker and colleagues.[3] These models are described in detail below.

Data

All data were annual state-level characteristics from 2001 through 2011 taken from multiple sources.[4] Each observation in the data represented a state for a given year between 2002 and 2011. Some factors were lagged 1 year to account for a deterrent effect and to impute data missing for a later year. Eleven states were excluded from the final analyses as potential controls because they had implemented policies early in the postintervention period or because they were missing data on a key covariate; they include Arkansas, California, Hawaii, Kentucky, Louisiana, Iowa, Illinois, Tennessee, Oklahoma, Oregon, Mississippi, and Florida.

Variables included in the analysis were similar to those controlled in other studies on the impact of precursors.[5]

Outcome variables: We modeled two outcome variables: the total lab seizure rate per 100,000 population and the small toxic lab seizure (STL) rate per 100,000 population. Small toxic labs are defined as labs with a capacity of 1 pound or less. Data on methamphetamine seizure incidents from the National Seizure System maintained by Drug Enforcement Administration's El Paso Intelligence Center (DEA EPIC) were aggregated to get the number of methamphetamine lab seizures per state per year.[6] The rates were computed using the Census annual population estimate as a denominator multiplied by 100,000. It is expressed as the rate per 100,000 people. The rates were transformed by taking the log base 10 to approximate the normal distribution required for a linear model.

Other factors were controlled in this model. The control variables included the following

- Client rate: The rate of substance abuse clients reported annually into the Substance Abuse and Mental Health Services Administration through the National Survey of Substance Abuse Treatment Services (N-SSATS) per 100,000 people. This factor is lagged 1 year to account for the possibility that the number of substance abuse clients has more of an impact on the future number of labs seized than the current number of labs. Lagging these data also allows us to make up for unavailable data in 2011. The client rate is not available for 2002. The 2001 value is used to impute that value.

- Region: Regional factors are expected to affect the methamphetamine problem and domestic production. We cannot identify or control for all of the potential factors that influence lab seizures for the region, so we include a set of dummy variables indicating the census division to approximate the potential influence of regional factors. Divisions include the following:
  1 = New England
  2 = Middle Atlantic
  3 = East North Central
  4 = West North Central
  5 = South Atlantic
  6 = East South Central
  7 = West South Central
  8 = Mountain
  9 = Pacific (referent category).
- Demographics: Some demographic groups are more likely to use methamphetamine than other groups. We controlled for the demographic composition of the state population to account for potential demand for the drug. The percentage of the population that is non-Hispanic white, male, Hispanic, and under age 18 were computed annually for each state from Census intercensal population estimates.
- Distance to Mexico: The approximate number of miles between the state and the nearest Mexican border city was taken from Cunningham et al. (2010)[8] . The number of miles were included as a set of categories with the farthest distance (1,800 miles) as the reference category. This variable attempts to account for the supply of imported methamphetamine on domestic production.
- Funding: The Community Oriented Policing Services (COPS) funding amount from DEA was adjusted to 2012 dollars using the Consumer Price Index and divided by 1,000 to adjust the scale of the dollar amounts. This variable controlled for law enforcement activity specific to methamphetamine lab cleanups. It also helped to adjust for a possible downward bias in the 2011 reporting because of a discontinuation of COPS funding for a portion of that year.
- Police: The presence of police was measured as the annual number of employed law enforcement officers as a percentage of the total

population. Police data came from the Uniform Crime Report (UCR) Law Enforcement Officers Killed in Action (LEOKA) data set. This factor was lagged 1 year to account for the possibility that the presence of police has a deterrent effect on the future number of labs seized. Lagging these data also allowed us to make up for unavailable data in 2011.

- Arrests: The drug arrest rate was measured as the number of drug arrests (UCR offense code 18) per 100,000 population. The data come from the Uniform Crime Reporting Program Data: Arrests by Age, Sex, and Race, Summarized Yearly. Data for Florida were not reported in this data set. This factor was lagged 1 year to account for the possibility that the number of drug arrests has a deterrent effect on the future number of labs seized. Lagging these data also allowed us to make up for unavailable data in 2011.

While recent analyses of methamphetamine precursor laws have used relatively similar parsimonious models, our model may still be underspecified. For example, we did not control for alternative drug use.

Difference-inDifferences Estimation Model

The DD model is a regression model that compares over time the outcomes for a unit of analysis that has been exposed to a treatment or intervention (referred to as a case) with the outcomes of at least one unit that has not been exposed to the treatment or intervention (referred to as a control). The case is exposed to the intervention at some point after the first period of time; the control is never exposed to the intervention during the course of the study. The impact of the intervention is represented by the difference in differences. In this case there are two differences. The first difference is between the average outcomes of the case and the control, respectively, in the post-intervention period and the preintervention period. The second difference subtracts the control difference between the two periods from the case difference. It can be written as equation 1.

EQ. 1: $DD = (\text{y-bar}_{Case,post-} - \text{y-bar}_{Case,pre-}) - (\text{y-bar}_{Control,post-} - \text{y-bar}_{Control,pre-})$

For a DD model, the data consist of one observation for each geographic unit, which is represented by subscript i and each unit of time which is represented by subscript j. In our analysis, each observation represents a state in each year from 2000 through 2010. Since the interventions were implemented in 2006, the preintervention period spans 2000 through 2006. The post period spans 2007 through 2010. A dummy-variable[9] indicating the postintervention period is specified; therefore, our DD model takes the form:

EQ. 2: $Y_{ij} = \beta_0 + \beta_1 \mathit{Post\text{-}Intervention\ Dummy} + \beta_2 \mathit{Oregon} + \beta_3 \mathit{Post\text{-}Intervention\ Dummy}^{*}\mathit{Oregon} + \beta_4 \mathit{Time} + \beta_5 X_{ij} + \varepsilon_{ij}$

Where Yij is the outcome for state i at period j; 130 is the intercept term; 131 is the coefficient on the postintervention period dummy variable; 132 is the coefficient on the Oregon dummy variable; 133 is the coefficient on the interaction term between Oregon and the postinteraction period dummy variable; 134 is the coefficient on the time variable; 135 is the set of coefficients on the set of control variables; and ε_{ij} is the error term.

This analysis primarily seeks to reliably estimate the coefficient 133. This estimate indicates the average amount of impact the prescription-only policy implemented in 2006 had over the period 2007 through 2010. This estimate will allow us to say the extent to which each policy contributed to the change in domestic methamphetamine lab seizures after controlling for other factors.[10]

DD estimation has some known limitations described in the academic literature. Beasley and Case (2000) describe the endogeniety of interventions, i.e., the fact that policies are made in response to the same conditions that lead to the outcome. Heckman (2000) and Bertrand and colleagues (2004) showed that because of serial correlation in the outcomes over time, difference-in-differences models tended to underestimate the standard error of the intervention coefficient and therefore overestimate the test statistic, leading to the interpretation of statistically significant differences between the case and control units. Abadie and colleagues (2010) argue that the selection of control units are made on the basis of subjective measures of affinity between case and control units and that there is uncertainty in the control units' ability to produce the counterfactual outcome trend that the case would have experienced if the intervention has not taken place. This is an additional source of uncertainty beyond that measured by the standard error. We attempt to address these limitations in the analysis.

To account for autocorrelation, we implemented this model using Generalized Estimating Equations (GEE) in SAS Proc Genmod with a repeated statement specifying the compound symmetry covariance structure to account for the autocorrelation in the covariates across time periods for each state. The covariance structure was determined by examining the working correlation matrix estimated when specifying an unstructured covariance structure and by comparing the quasi-likelihood indicator criteria (QIC) statistics for models specifying five different covariance structures: independence, compound symmetry, first-order autocorrelation, unstructured, and 1-dependent. The unstructured covariance structure allows for correlations to be different in each comparison of times without any specific pattern. The unstructured working correlation matrix indicated high constant correlation over time. Since the correlations seem constant, a compound symmetry structure is more appropriate.

The QIC values for the GEE models were similar with the independent, compound symmetry, and autocorrelation structures specified, but the QIC was usually lowest for the independence structure with autoregressive next. This indicates that those structures fit the model better. Independence in the measures across time is not a logical assumption given the nature of the data, and the structure of the correlation matrix specified by the unstructured covariance structure does not show declining correlations over time described in the autoregressive structure. The QIC supports our choice of a compound symmetry covariance structure.

Synthetic Control Method

We validated our model findings using the synthetic control method. The synthetic control method introduced by Abadie and colleagues (2010) is a modification on the DD method that creates a data-driven synthetic control that represents the counterfactual of the case in the absence of the intervention. The synthetic control method has two advantages. It allows for transparency and objectivity in the selection of control. It also safeguards against extrapolation of the counterfactual by creating the synthetic control to match the case closely in the pre-intervention period.

We implemented the synthetic control method in Stata using the synth ado program.[11] The program uses the set control states to create a synthetic form of the case study state by weighting the control states. The treated and synthetic control states are matched on the outcome and any combination of covariates in the preintervention period so that the mean squared error of the prediction variables is minimized. Then the model interpolates the trajectory of the synthetic state over the postintervention period assuming that the intervention was not implemented. In preliminary analyses, we tested the robustness of the model matching the state and synthetic controls on the outcomes alone and on the outcomes and all covariates controlled in the GEE models. All results presented here are based on a model matching on the outcome and most covariates controlled in the GEE models.[12] The synthetic control method does not generate a simple test statistic to determine whether the difference between the case study and synthetic control state is statistically significant. To test whether the results are likely to be found by chance in Oregon, we ran the model assigning Oregon's neighboring states that met our criteria for inclusion as controls (Washington, Idaho, and Nevada) as the case study state and allowed the model to generate a synthetic control to compare what would have happened relative to the experience in each of those states. If the results were found to be similar to Oregon's, then we could not dismiss the possibility that our findings for Oregon were due to chance.

Results

Prescription-only had significant impacts on lab seizure rates compared with a selected group of controls. Contrary to the findings in Cunningham et al. (2012)[13] and Strauberg and Sharma (2012)[14] , our analysis found that lab seizure rate fell by more than 90 percent in Oregon after the prescription-only requirement was implemented after adjusting for other factors. While 90 percent seems very high, the estimate should be considered in the context that the rate has been declining and was relatively low before the policy was implemented. The impact of the prescription-only requirement was validated when the case study state was compared with an empirically generated synthetic control. The synthetic control method confirmed the direction of the impact in Oregon. Our placebo analysis that assigned Oregon's neighbor states as the control state showed that the reductions seen in Oregon were not projected in those states, giving some indication that the Oregon reduction was not found by chance. We cannot determine the extent of the impact using the synthetic control method because of the poor fit of the model in the period prior to the policy's implementation.

Our analysis differs from the two recent studies cited above in the methodology, including the analytical approach and model specification, and the date on which the incident data were pulled.[15]

Interpretation of Results

The key finding from the GEE model is the coefficient on the interaction between the case study state and the postintervention period indicators. Since the outcome data were transformed to improve the model fit, we back-transformed the coefficients for ease of interpretation.[16] Four estimates are presented in Table 5. They represent the model specifications. Each group of covariates was modeled on the two outcomes described above: the lab seizure rate including all capacities and the small toxic lab seizure rate. The unadjusted model adjusts only for the policy, state, time effects, and the interaction between the case study state and the postintervention period indicators. The adjusted model adjusts for those factors and controls all covariates described above. The unadjusted impacts are interpreted as the percent change in the rate resulting from the implementation of the policy adjusting only for temporal factors. Adjusted factors are interpreted as the percent change in the rate resulting from the implementation of the requirement after controlling for other factors that may also affect the change in the seizure rate. Impacts are determined to be statistically significant if the p-value is less than 0.05.

Table 5. Methamphetamine PSE Restriction Policy Impact on Lab Seizure Rates per 100,000 Population Estimated from Difference-in- Difference

	Unadjusted				Adjusted for other factors			
	All lab seizures		Small toxic labs		All lab seizures		Small toxic labs	
Oregon	Policyimpact	P-value	Policyimpact	P-value	Policyimpact	P-value	Policyimpact	P-value
prescription-only	-98.72	0	-98.53	0	-97.40	0	-97.01	0

Source: GAO analysis.

Note: Statistical significance is indicated by a p-value of 0.05 or lower.

The key finding from the synthetic control model is the difference in the estimated lab seizure rate in the years after 2006 between the case study state and the synthetic control. Differences in the postintervention period can be attributed to the impact of the policy when the two match closely in the preintervention period. Since the states did not always have a close match in the preintervention period and the model does not generate a test statistic to indicate whether the differences between the case study and synthetic control are statistically significant, we do not present numerical results indicating the size of the impact of the policy from this analysis; instead we used the results to validate the direction of the findings of the GEE models.

End Notes

[1] Medical and psychological effects of meth abuse can include aggression, memory loss, heart damage, hyperthermia, and psychotic behavior. While legitimately manufactured meth can be used to treat such medical conditions as narcolepsy, attention deficit disorder, obesity, and depression, alternative drugs are more frequently used for treatment of these conditions.

[2] Illicit meth can be manufactured in various forms and can be smoked, snorted, injected, and taken orally. Super labs are capable of producing over 10 pounds of meth in a 24- hour period. According to estimates by DEA's Office of Diversion Control, meth produced by Mexican drug-trafficking organizations makes up approximately 80 percent of the illicit meth consumed in the United States. Small toxic labs produce less than 2 ounces (56 grams) of finished product. Most labs found in the United States are small toxic labs.

[3] Ephedrine can be used as a substitute for PSE when making methamphetamine.

[4] Meth lab incidents include seizures of labs, dump sites, chemicals, and glassware. Law enforcement agencies report meth lab incidents to the National Seizure System (NSS) maintained by DEA's El Paso Intelligence Center (EPIC). EPIC monitors and tracks all meth-related information nationally and internationally that is reported to the NSS.

[5] Pub. L. No. 109-177, 120 Stat. 256 (2006). The CMEA also placed restrictions on the precursor chemicals ephedrine and phenylpropanolamine. At the time, the industry had removed these drugs from the market for safety reasons.

[6] Oregon, Mississippi, and select Missouri cities and counties have passed laws or ordinances requiring individuals to obtain a prescription from a health care provider in order to purchase PSE products.

[7] According to DEA officials, records of incidents can be updated within the NSS or new records added as new data and information becomes available or is submitted. Consequently, the number of total number of incidents may vary over time. The data analyzed for this review were pulled from the NSS on October 1, 2012.

[8] The states we selected as case study states included the electronic tracking states of Iowa, Kentucky, Missouri, and Tennessee, and the prescription-only states of Mississippi and Oregon. Select Missouri cities and counties have passed laws or ordinances requiring individuals to obtain a prescription from a health care provider in order to purchase PSE products.

[9] These border states included Alabama, Arkansas, California, Idaho, Louisiana, Nevada, Tennessee, and Washington.

[10] IMSHealth Inc. is a provider of consulting and analytical information and services for the health care industry.

[11] CHPA is a trade association that represents U.S. manufacturers of nonprescription medications.

[12] The Birch reduction method-also known as the Nazi method-uses ephedrine or pseudoephedrine, anhydrous ammonia, and sodium or lithium metal and requires less than 1 hour to produce methamphetamine that is about 95 percent pure. The Red Phosphorus method-also known as the Red P method- uses ephedrine or pseudoephedrine, iodine or hydriodic acid, and red phosphorus and takes over 3 hours to produce methamphetamine that is approximately 90 percent pure.

[13] This method involves the synthesis of methylamine and phenylacetone also known as 1-phenyl-2- propanone or P-2-P. Motorcycle gangs used this method to manufacture meth in the 1960s and 1970s.

[14] The CMEA was enacted in the USA PATRIOT Improvement and Reauthorization Act of 2005, Pub. L. No. 109-177, 120 Stat. 192 (March 9, 2006). The CMEA also set limits on imports of meth precursor chemicals, increased penalties for meth production and trafficking, and required the Department of State to work with Mexico to help prevent the smuggling of illicit meth across the U.S.-Mexico border.

[15] Sellers must enter the name of the product and quantity sold into the logbook. Customers must write or enter into the logbook their name, address, date, and time of sale and also sign the logbook. Sellers must verify that the customer's name matches the name written in the logbook by that individual and that the date and time of sale are correct.

[16] For example, Alaska, Iowa, Indiana, Minnesota, and Wisconsin have implemented lower monthly sales limits than the 9-gram limit established by the CMEA.

[17] These systems generally allow retailers to override a stop sale alert if certain conditions are met (i.e., under a threat of violence). Such sales are known as exceedances.

[18] The National Association of Drug Diversion Investigators distributes the NPLEx system to participating states that pass a law requiring that all sales of PSE be tracked electronically in real time and illegal transactions be denied.

[19] This analysis excluded incident reports that did not include information on lab capacity or type.

[20] Paul A. Blostein; Brian R. Plaisier, MD; Sheldon R. Maltz, MD; Scott B. Davidson, MD; Eric W. Wideman, DO; Eric C. Feucht, MD; and Sheri L. VandenBerg, RN. "Methamphetamine Production is Hazardous to Your Health," *Journal of Trauma, Injury, Infection, and Critical Care*, vol. 66, no.6, (2009).

[21] According to the director of the Vanderbilt University Regional Burn Center, on the basis of his experience, meth lab-related burn victims generally try to conceal the cause of their injuries and health providers are hesitant to determine whether the injuries were due to illicit meth production because many insurers will not pay for such care under those circumstances.

[22] As part of reporting a lab seizure to the DEA's NSS, law enforcement is required to report on the number of children affected by the lab, such as those living at the site as well as those that might have visited the site.

[23] According to information provided by Missouri officials, the average child remains in state care for 369 days at a cost of $18.35 per day.

[24] According to Division of Family Services officials, they now try to locate family members that are clean of addiction and crime who can care for these children.

[25] Tennessee officials report that a child spends an average of 399.5 days in foster care, with an overall average cost of $108 dollars per day to the state.

[26] Resource Conservation and Recovery Act of 1976, Pub. L. No. 94-580, 90 Stat. 2795.

[27] Some states, such as South Carolina, California, and Missouri, have established their own independent state-run cleanup programs.

[28] In 2004, 98 percent of the laboratories seized were producing meth. While most clandestine labs produce meth, some labs produce other substances such as Ecstasy (MDMA). The number of substances that are manufactured is constantly growing as new controlled substance analogs are developed to circumvent controlled substance laws.

[29] For example, decontamination of an average-sized site has been estimated to cost around $50,000.

[30] 29 C.F.R. § 1910.120

[31] With such systems in place, law enforcement no longer would need to go from retailer to retailer to examine or make copies of written logs to look for investigative leads.

[32] The system can notify law enforcement when a particular individual purchases or attempts to purchase a PSE product. Law enforcement can also run specific reports on individuals or pharmacies. For example, a store report provides law enforcement with transaction summaries, as well as transaction and compliance details. An activity report can provide information on a specific person's ID, specific pharmacy, and activity type (purchase, attempt, block, return, or inquiry).Two of the 17 states using NPLEx (Tennessee and Indiana) also use their own tracking systems in addition to NPLEx.

[33] Lead-generating systems report sales information to a database that can be accessed by law enforcement to identify purchase limit violations or help generate leads on potential diversion for meth production.

[34] With regard to the implementation of electronic tracking systems, some law enforcement officials have raised concerns specifically about the NPLEx system related to accessing sales data, among other issues. In November 2012, Senator Ron Wyden of Oregon and Representative Phil Roe of Tennessee requested that DOJ conduct an investigation into the NPLEx system and its operation. According to the Deputy Assistant Administrator for DEA's Office of Diversion control, DOJ is reviewing this request.

[35] U.S. Department of Justice. National Drug Intelligence Center, National Drug Threat Assessment 2011. (Washington, D.C.: August 2011).

[36] Boxes of PSE purchased for $7 or less can be sold for between $30 and $100.

[37] NMPI is a national High Intensity Drug Trafficking Area (HIDTA) program initiative funded by the ONDCP through the Southwest Border HIDTA California Region. The NMPI is a national strategy, intelligence-sharing, and training initiative addressing methamphetamine and pharmaceutical drug crimes in the United States. The mission of NMPI is to reduce the availability of methamphetamine and its precursor chemicals throughout the United States. NMPI also seeks to reduce pharmaceutical drug crimes by utilizing best practices for investigations and intelligence collection and analysis. The NMPI has a National Advisory Board consisting of four federal and six state and local representatives from various regions of the United States. The purpose of the HIDTA program (created by the Anti-Drug Abuse Act of 1988, Pub. L. No. 100-690, 102 Stat. 4181) is to reduce drug trafficking and production in the United States by, among other things, facilitating cooperation among federal, state, local, and tribal law enforcement agencies. There are currently 28 HIDTA regions, which include approximately 16 percent of all counties in the United States and 60 percent of the U.S. population. HIDTAdesignated counties are located in 46 states, as well as in Puerto Rico, the U.S. Virgin Islands, and the District of Columbia. The HIDTAs are directed and guided by executive boards composed of an equal number of regional federal and non-federal (state, local, and tribal) law enforcement leaders.

[38] 2011 Ark. Acts 588. See Ark. Code Ann. §§ 5-64-1103 to -1105.

[39] Ark. Code Ann. § 5-64-1103.

[40] The analysis controlled for other factors that could have an impact on the number of meth lab incidents in a given state such as region of the country, ethnic composition, the proportion of the state population that is male, distance from the Mexican border to account for international supply, police presence, and the state drug arrest rate, among others. Consistent with completed peer-reviewed research, the analysis was run on incidents of all lab seizures of all productive capacities, as well as small toxic labs. For more information on the methodology used for this model and the results of this analysis, see appendix III.

[41] Because of the lack of sufficient data from the time Mississippi adopted the prescription-only approach in July 2010, we could not perform a similar analysis for Mississippi.

[42] Contrary to the findings in Cunningham et al. (2012) and Strauberg and Sharma (2012), our analysis found that the lab seizure rate fell significantly in Oregon after the prescription-only policy was implemented after adjusting for other factors. The differences in our findings are likely due to differences in model specification, analytical approach, and date we pulled the data from the NSS for analysis. See appendix III for more details on our methodology.

[43] The only state Oregon shares a border with that has adopted electronic tracking during the period of this analysis was Washington, which implemented the NPLEx system on October 15, 2011. In addition, some retailers located in the neighboring states of California, Idaho, and Nevada, which have not adopted electronic tracking or the NPLEx system, use the software platform the NPLEx system is based on to track PSE sales within their stores and meet logbook requirements. Known as MethCheck, the software platform was developed by Appriss, Inc., a government technology provider based in Louisville, Kentucky. It is a real-time electronic reporting system that allows pharmacy employees to view each customer's purchasing history at the point of sale and makes that information available to law enforcement.

[44] Three states bordering Mississippi (Louisiana, Alabama, and Tennessee) have adopted the electronic tracking approach for restricting PSE product sales and have implemented the NPLEx electronic tracking system. These states implemented the NPLEx system at different times: Alabama in January 2011, Louisiana in June 2010, and Tennessee in January 2012. Arkansas also borders Mississippi and has adopted the electronic tracking approach but has implemented its own electronic tracking system.

[45] Because these states adopted the system at about the same time or after Mississippi adopted its prescription-only approach, data are not available to determine the extent to which the volume of PSE sales-related activities involving individuals using Mississippi identifications changed over time.

[46] 2011 Ark. Acts 588. See Ark. Code Ann. §§ 5-64-1103 to -1105.

[47] 2012 Ala. Acts 2012-237. See Ala. Code § 20-2-190(d).

[48] Because Walmart does not report to outside parties the volume of sales in its stores of individual products, the sales data for PSE products do not include the PSE products sold at Walmart stores. Consequently, if sales at Walmart stores were to be included, the total volume of all sales for PSE products would likely be higher than reported by this estimate. Data on prescriptions for PSE are tracked through the Mississippi Prescription Drug Monitoring Program, which includes data from all licensed pharmacies in Mississippi.

[49] Oregon did not initiate its prescription drug monitoring program until 2011. From October 2011 to November 2012, Oregon officials report that about 140,000 prescriptions for PSE have been filled.

[50] The FDA Nonprescription Drug Advisory Committee held a hearing on December 14, 2007, in response to a citizen petition submitted to FDA that asserted that the available data do not

support the adult and pediatric doses of phenylephrine hydrochloride and phenylephrine bitartrate that are generally recognized as safe and effective in the OTC drug monograph as set forth in 21 C.F.R. part 341. The citizen petitioners included Leslie Hendeles, PharmD, Professor, Pharmacy and Pediatrics, University of Florida; Randy C. Hatton, PharmD FCCP BCPS, Co-Director, Drug Information and Pharmacy Resource Center, Clinical Professor, University of Florida; and Almut G. Winterstein, PhD, Assistant Professor, Department of Pharmacy Healthcare Administration, University of Florida.

[51] According to FDA procedures and processes for making changes to approved OTC monographs, including those involving changes in dosage levels, additional studies of the effectiveness of higher dosages of PE would need to be conducted by industry or citizen petitioners. We contacted the citizen petitioners, who indicated they were not aware of any additional study of the issue since the advisory committee meeting in 2007.

[52] In its briefing to the FDA advisory committee, CHPA stated that its assessment of available evidence at the time supports the conclusion that PE at dosages of 10 milligram is safe and effective as a nasal decongestant for over-the-counter use in adults. However, CHPA has also noted limitations of PE that make it a less than direct substitute for PSE in some applications. For example, PE is metabolized more quickly than PSE and is thus available only in conventional formulations to be taken every 4 hours, whereas PSE can be made available as 12-hour and 24-hour sustained formulations, a fact that is said to increase consumer convenience and reduce the risk of medication errors. Additionally, CHPA notes that as consumers respond differently to different ingredients, many choose PSE because it works better for them.

[53] Although it is possible that some Medicaid plan participants previously purchased PSE products on their own and at their own expense prior to the law being implemented but now would be required to obtain a prescription, the officials stated that the data are not available to quantify the number of patients this is the case for. To the extent that these participants are now having Medicaid cover their PSE prescriptions, there is likely an increase in cost to the program.

End Notes for Appendix I

[1] According to DEA officials, records of incidents can be updated within the NSS or new records added as new data and information becomes available or is submitted. Consequently, the number of total number of incidents may vary over time. The data analyzed for this review were pulled from the NSS on October 1, 2012.

[2] The dates these states joined NPLEx were: Alabama, January 1, 2011; Louisiana, June 15, 2010; and Tennessee, January 1, 2012.

End Notes for Appendix III

[1] Bruce Meyer, "Natural and Quasi-Natural Experiments in Economics" *Journal of Business and Economic Statistics*. JES Symposium on Program and Policy Evaluation, .vol.13(2) (2000):151-161.

[2] Alberto Abadies, Alexis Diamond, and Jen Hainmueller, "Synthetic Control Methods for Comparison Case Studies: Estimaing the Effect of California's Tobacco Control Program." *American Statistical Association*. June 2010. vol. 105. (490): 493-505.

[3] James Nonnemaker, Mark Engelen, and Daniel Shive, "Are Methamphetamine Precursor Control Laws Effective Tools to Fight the Methamphetamine Epidemic?" *Health Economics*. vol. 20 (2000):519-531.

[4] We would like to acknowledge Monica Wade and Creston Dalmadge at Winston-Salem State University for sharing their dataset with us for the preliminary analyses. All data used for this report come from the original sources.

[5] For examples of peer-reviewed articles studying the impact of methamphetamine precursor restrictions, see James K. Cunningham, Jane Carlisle Maxwell, Octavio Campollo, Kathryn I. Cunningham, Lon-Mu Liu, and Hui-Lin Lin. "Proximity to the US– Mexico border: A Key to Explaining Geographic Variation in US Methamphetamine, Cocaine and Heroin Purity," *Addiction* 105 (2010): 1785-1798; James K. Cunningham, , Russell C. Callaghan, Daoqin Tong, Lon-Mu Liu, Hsiao-Yun Li, and William J. Lattyak. "Changing Over-the-Counter Ephedrine and Pseudoephedrine Products to Prescription Only: Impacts on Methamphetamine Clandestine Laboratory Seizures," *Drug and Alcohol Dependence* 126 (2012): 55-64; James K. Cunningham, Lon-Mu Liu, and Myra Muramoto. "Methamphetamine Suppression and Route of Administration: Precursor Regulation Impacts on Snorting, Smoking, Swallowing and Injecting," *Addiction* 103 , (2008): 1174-1186; and James Nonnemaker, Mark Engelen, and Daniel Shive, "Are Methamphetamine Precursor Control Laws Effective Tools to Fight the Methamphetamine Epidemic?" *Health Economics* 20 (2010):519-531.

[6] Methamphetamine seizure incidents were not completely captured by the NSS. Throughout 2012, EPIC added more than 30,000 new incidents from the HAZARD system into NSS. These incidents occurred in the period 1998 through 2012. The data for our analysis was extracted on Oct. 10, 2012. These data reflect all but 824 incidents that had not been reconciled between the two systems.

[7] The rate of treatment admissions from the Treatment Episode Data Set (TEDS) was considered as an additional control, but we decided to exclude it because of many missing data points in the series.

[8] See figure 1 on page 1787.

[9] Dummy variables are also called indicator variables. They are used to indicate the occurrence or presence of a concept. They take the value of 1 when the thing occurs and the value of 0 otherwise.

[10] The control variables are described in the data section of this document.

[11] Alberto Abadie, Alexis Diamond, and Jens Hainmueller. "Synth: An R Package for Synthetic Control Methods in Comparison Case Studies." *Journal of Statistical Software.* vol 42(13) (2011):1-17. We would like to acknowledge Jens Hainmeuller who assisted us in our verification of the Stata program.

[12] The regional division was not included as a covariate for matching to allow the model to select controls from across the country. The arrest rate was not included as a covariate because the pattern of missing data across the series made that variable incompatible with the Stata program.

[13] James K. Cunningham, Russell C. Callaghan, Daoqin Tong, Lon-Mu Liu, Hsiao-Yun Li, and William J. Lattyak. "Changing Over-the-Counter Ephedrine and Pseudoephedrine Products to Prescription Only: Impacts on Methamphetamine Clandestine Laboratory Seizures," *Drug and Alcohol Dependence* 126 (2012): 55-64.

[14] Christopher Stomberg and Arun Sharma, *Making Cold Medicine Rx-Only Did not Reduce Meth Use.* (Portland, OR: Cascade Policy Institute, February 2012).

[15] Our data were pulled in October 2012, following the publication of these reports. Important changes in the number of seizures covered in the data are described in the data section. Cunningham et al. used an Autoregressive Integrated Moving Average (ARIMA) with quasi control by comparing the trend in monthly seizure counts in Oregon and Mississippi compared to seizure counts in their neighbor states.(See Cunningham, Callaghan, Tong, Liu, Li, and Lattyak. "Changing Over-the-Counter Ephedrine and Pseudoephedrine Products to Prescription Only" 55-64.). Stromberg and Sharma estimated a DD model, but specify it with the outcome variable measured as an index and control only regional and temporal trends. (See Stomberg and Arun Sharma, *Making Cold Medicine Rx-Only Did not Reduce Meth Use.*

[16] Back transformation is preferred for precise interpretation of the results. See GAO, *Women's Pay: Gender Pay Gap in the Federal Workforce Narrows as Differences in Occupation, Education, and Experience Diminish,* GAO-09-279 (Washington, D.C.: March 17, 2009). Since we used the log base 10 transformation, the back transformation is 10beta instead of ebeta, which is denoted in the GAO report referenced here.

In: Methamphetamine
Editor: Ace M. Donijan

ISBN: 978-1-62618-497-8

Chapter 2

METHAMPHETAMINE*

Drug Enforcement Administration, Office of Diversion Control

TRADE NAME: DESOXYN®; STREET NAMES: METH, SPEED, CRYSTAL, GLASS, ICE, CRANK, YABA

Introduction

Methamphetamine is a highly addictive drug with potent central nervous system (CNS) stimulant properties. In the 1960s, methamphetamine pharmaceutical products were widely available and extensively diverted and abused. The placement of methamphetamine into schedule II of the Controlled Substance Act (CSA) in 1971 and the removal of methamphetamine injectable formulations from the United States market, combined with a better appreciation for its high abuse potential, led to a drastic reduction in the abuse of this drug. However, a resurgence of methamphetamine abuse occurred in the 1980s and it is currently considered a major drug of abuse.

* This fact sheet was released July 2012, by Drug Enforcement Administration, Office of Diversion Control.

The widespread availability of methamphetamine today is largely fueled by illicit production in large and small clandestine laboratories throughout the United States and illegal production and importation from Mexico. In some areas of the country methamphetamine abuse has outpaced heroin and cocaine.

Licit Uses

Methamphetamine was originally used in nasal decongestants and bronchial inhalers (the levo isomer of methamphetamine is still utilized for these indications). Later it was available in tablets and injectable formulations and used for weight control, depression, and to increase alertness and prevent sleep. A broad segment of society used methamphetamine products for stimulant effects. Today there is only one product, Desoxyn®, currently marketed in 5 mg tablets. Desoxyn® has very limited use in the treatment of obesity, and attention deficit hyperactivity disorder.

IMS Health® reports 15,000 prescriptions for methamphetamine were dispensed in the U.S. in 2011. DEA's 2012 Aggregate production quota for methamphetamine was established at 3,130 kilograms.

Chemistry/Pharmacology

Methamphetamine is chemically and pharmacologically similar to amphetamine although it has more potent effects on the CNS that can last for 6 to 8 hours. Methamphetamine increases the release of the neurotransmitter, dopamine, which stimulates brain cells, enhancing mood and energy. At low doses, methamphetamine produces such effects as increased wakefulness, increased physical activity, increased heart rate and blood pressure, decreased appetite, increased respiration and body temperature (hyperthermia), and euphoria. High-dose chronic use has been associated with irritability, tremors, convulsions, anxiety, paranoia, and neurotoxic effects that cause damage to neurons and blood vessels.

Aggressive and violent behavior, often directed at spouses and children, pose a significant risk to those individuals in contact with methamphetamine addicts. Death has resulted from extreme anorexia, hyperthermia, convulsions, and cardiovascular collapse (including stroke and heart attacks).

Illicit Use

Methamphetamine is abused for its stimulant and euphoric effects. It can be taken orally, snorted, smoked, and injected. Smoking or injecting methamphetamine results in intense euphoria and is often associated with binge use, large escalation in dose with rapid tissue tolerance, and high rates of dependence and addiction. "Ice," "Glass," and "Crystal" are all terms for concentrated *d*-methamphetamine HCl chunks that are smoked. Yaba is a Thai name for a colored tablet containing methamphetamine combined with caffeine which is gaining popularity among individuals who frequent "raves." According to the 2010 National Survey on Drug Use and Health (NSDUH), 13.01 million individuals, aged 12 and older, reported nonmedical use of methamphetamine at least once in their lifetime. NSDUH reports that there were 105,000 people aged 12 and older who used methamphetamine for the first time in 2010. The Drug Abuse Warning Network (DAWN ED), reports 64,117 emergency department visits associated with methamphetamine in 2009. The 2011 Monitoring the Future survey indicates that 1.3% of 8th graders, 2.1% of 10th graders, and 2.1% of 12th graders used methamphetamine in their lifetime. In 2010, the American Association of Poison Control Centers reported 3,012 total poison exposures and 14 deaths related to methamphetamine. In 2009, there were 2,183 methamphetamine total exposures and seven deaths.

Illicit Production and Distribution

Mexican drug trafficking organizations have become the primary manufacturers and distributors of methamphetamine to cities in the Midwest and West. These criminal organizations are able to supply large amounts of methamphetamine at high purity and low cost. Domestic independent laboratory operators also produce and distribute methamphetamine but usually on a smaller scale.

Of particular concern is the toxic waste associated with these labs and the fact that many individuals, including children, are at risk of exposure to these toxic chemicals.

DEA's National Forensic Laboratory Information System (NFLIS) and System to Retrieve Information from Drug Evidence (STRIDE) data indicate that law enforcement officials submitted 148,375 exhibits, 160,095 exhibits, and 143,478 exhibits identified as methamphetamine to federal, state and local forensic laboratories in 2009, 2010 and 2011, respectively.

There are many ways to manufacture methamphetamine. The methods used are directly impacted by the availability of precursor chemicals and ease of synthesis. Drug traffickers are continually looking for loopholes in chemical control regulations and altering their methods of synthesis in order to continue their illegal activity.

Currently, methamphetamine is primarily produced by utilizing diverted pseudoephedrine combination products. The Combat Methamphetamine Epidemic Act of 2005 requires retailers of non-prescription products containing pseudoephedrine, ephedrine and phenylpropanolamine to place these products behind the counter or in a locked cabinet. Consumers must show identification and sign a logbook for each purchase.

Control Status

Methamphetamine is in schedule II of the CSA.

In: Methamphetamine
Editor: Ace M. Donijan

ISBN: 978-1-62618-497-8

Chapter 3

METHAMPHETAMINE ADDICTION: PROGRESS, BUT NEED TO REMAIN VIGILANT*

National Institute on Drug Abuse

DECLINES IN METHAMPHETAMINE ABUSE BY YOUTH

- Approximately 13 million people 12 years and older have abused methamphetamine in their lifetimes; in 2010, approximately 353,000 were current users (NSDUH).
- According to NIDA's 2010 Monitoring the Future Survey, the abuse rate among 8th 10th, and 12th graders declined significantly between 1999 and 2007 and has remained unchanged since then (Figure 1).
- Abuse remains noteworthy in certain areas of the country with indicators suggesting particular problems in Hawaii, the West Coast, and the Midwest.
- Methamphetamine's potent addiction liability and destructive health and social consequences make its abuse particularly dangerous.

Methamphetamine acts by increasing the release of dopamine in the brain, which leads to feelings of euphoria. However, this surge of pleasure is followed by a "crash" that often leads to repeated use of the drug and

---

* This research update was released by the National Institute on Drug Abuse, November 2011.

eventually to difficulty feeling any pleasure at all, especially from natural rewards.

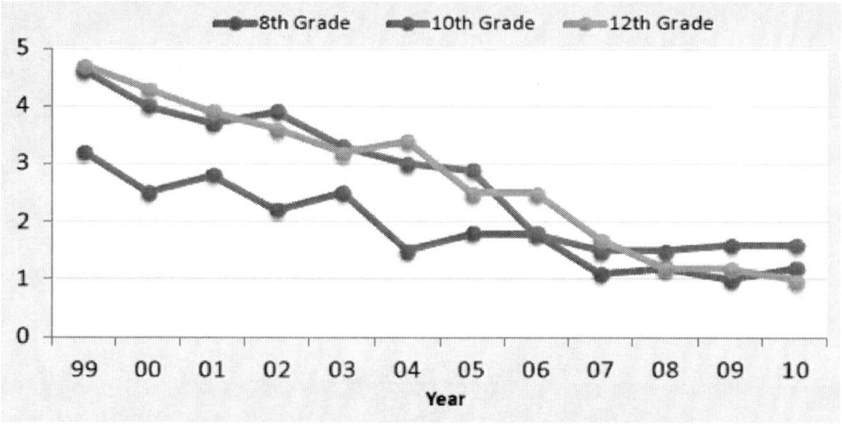

Source: University of Michigan, 2010 Monitoring the Future Study.

Figure 1. Percent of Students Reporting Past Year Use of Methamphetamine, by Grade.

Long-term methamphetamine abuse also results in many damaging physical and psychiatric effects, such as:

- Addiction
- Violent Behavior
- Anxiety
- Confusion
- Insomnia
- Psychotic symptoms (e.g. paranoia, hallucinations, delusions)
- Cardiovascular problems (e.g. rapid heart rate, irregular heartbeat, increased blood pressure, stroke).

WHAT DOES METHAMPHETAMINE DO TO THE BRAIN?

Methamphetamine's adverse effects on the brain are clear. Imaging studies have demonstrated changes in the dopamine system (important for reward, motivation, and learning) as well as structural and functional deficits in brain areas associated with emotion and memory. These may account for the psychiatric and cognitive problems observed in chronic abusers. Fortunately, some of the methamphetamine-induced deficits in dopamine function have been shown to recover, at least partially, with extended abstinence (Figure 2). But even with the partial recovery found in some brain regions following protracted abstinence, other regions do not show recovery of function—suggesting that long-lasting and even permanent brain changes may result from methamphetamine abuse.

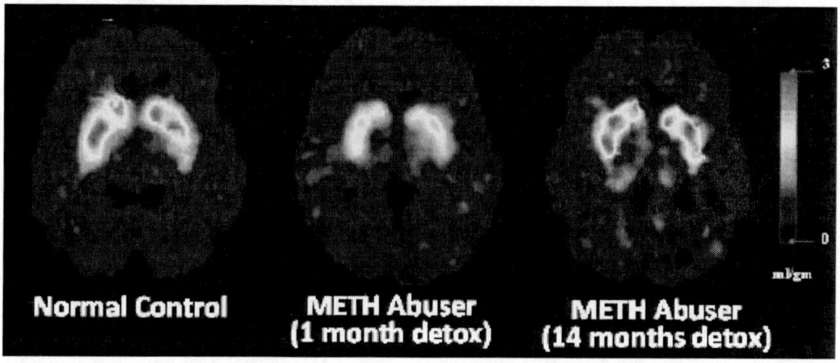

Source: Volkov, ND et al. Journal of Neuroscience 21, 9414-9418, 2001.

Figure 2. Patial Recovery of Brain Dopamine Transporters in Methamphetamine (METH) Abuser After Protracted Abstinence.

METHAMPHETAMINE AND HIV

Methamphetamine is inextricably linked with HIV, hepatitis C, and other sexually transmitted diseases. Its abuse increases the risk of contracting HIV not only through the use of contaminated injection equipment, but also through increased risky sexual behaviors and through physiological changes that may favor HIV transmission.

Methamphetamine abuse may also affect HIV disease progression. For example, clinical studies suggest that current methamphetamine abusers on highly active antiretroviral therapy may be at greater risk of developing AIDS than non-users, possibly due to poor medication adherence or interactions between methamphetamine and HIV medications. Similarly, preliminary studies suggest that interactions between methamphetamine and HIV itself may lead to more severe consequences for HIV-positive patients who abuse methamphetamine, including greater brain damage and cognitive impairment. More research is needed to better understand these interactions.

TREATMENTS FOR METHAMPHETAMINE ADDICTION

Methamphetamine addiction can be successfully treated. For example, the Matrix Model consists of a 16-week intervention that includes intensive group and individual therapy to promote the behavioral changes needed to remain abstinent, prevent relapse, and establish a lifestyle unrelated to drugs. When applied to methamphetamine abusers, the Matrix Model has been shown to significantly reduce drug use (Figure 3).

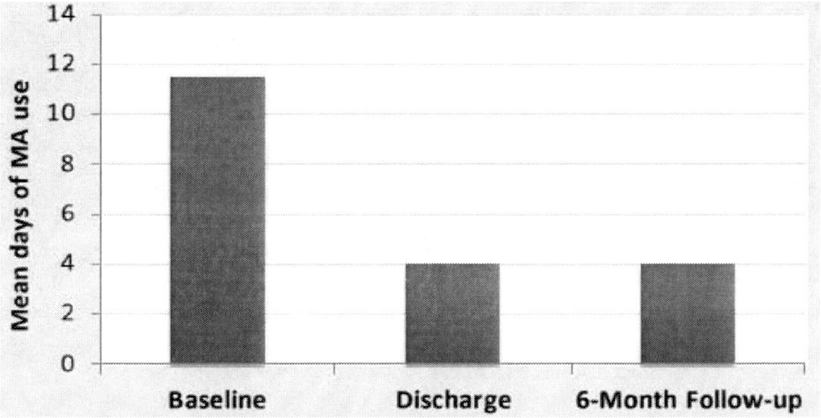

Source: Rawson et al. (2004) Addiction 99: 708-17.

Figure 3. Efficacy of the Matrix Model.

Motivational Incentives for Enhancing Drug Abuse Recovery (MIEDAR), an incentive-based method for promoting cocaine and methamphetamine abstinence, is another treatment that has demonstrated efficacy in methamphetamine abusers through NIDA's National Drug Abuse Clinical Trials Network.

There are currently no medications approved for the treatment of methamphetamine addiction. However, NIDA has made medications development a high priority. For example, a recent clinical trial revealed that the antidepressant bupropion, marketed as Welbutrin® is effective in reducing methamphetamine abuse in low/moderate users. Additional clinical trials of bupropion, and other medications (e.g., the antidepressant Mirtazapine) are currently underway. NIDA is also pursuing the novel approach of using vaccines, where antibodies are generated against the drug itself, to treat methamphetamine addiction. The antibodies would bind methamphetamine while it is still in the bloodstream, reduce its entry into the brain, and block its rewarding effects.

For further information please visit
NIDA on the web at www.drugabuse.gov or contact:
Public Information and Liaison Branch
Office of Science Policy and Communications
Phone 301-443-1124/Fax 301-443-7397
information@nida.nih.gov

In: Methamphetamine
Editor: Ace M. Donijan

ISBN: 978-1-62618-497-8

Chapter 4

NIDA INFOFACTS: METHAMPHETAMINE*

National Institute on Drug Abuse

Methamphetamine is a central nervous system stimulant drug that is similar in structure to amphetamine. Due to its high potential for abuse, methamphetamine is classified as a Schedule II drug and is available only through a prescription that cannot be refilled. Although methamphetamine can be prescribed by a doctor, its medical uses are limited, and the doses that are prescribed are much lower than those typically abused.

Most of the methamphetamine abused in this country comes from foreign or domestic superlabs, although it can also be made in small, illegal laboratories, where its production endangers the people in the labs, neighbors, and the environment.

HOW IS METHAMPHETAMINE ABUSED?

Methamphetamine is a white, odorless, bitter-tasting crystalline powder that easily dissolves in water or alcohol and is taken orally, intranasally (snorting the powder), by needle injection, or by smoking.

* This is an edited, reformatted and augmented version of a National Institute on Drug Abuse, National Institutes of Health publication, dated March 2010.

How Does Methamphetamine Affect the Brain?

Methamphetamine increases the release and blocks the reuptake of the brain chemical (or neurotransmitter) dopamine, leading to high levels of the chemical in the brain—a common mechanism of action for most drugs of abuse. Dopamine is involved in reward, motivation, the experience of pleasure, and motor function. Methamphetamine's ability to release dopamine rapidly in reward regions of the brain produces the intense euphoria, or "rush," that many users feel after snorting, smoking, or injecting the drug.

Chronic methamphetamine abuse significantly changes how the brain functions. Noninvasive human brain imaging studies have shown alterations in the activity of the dopamine system that are associated with reduced motor skills and impaired verbal learning.[1] Recent studies in chronic methamphetamine abusers have also revealed severe structural and functional changes in areas of the brain associated with emotion and memory,[2,3] which may account for many of the emotional and cognitive problems observed in chronic methamphetamine abusers.

Repeated methamphetamine abuse can also lead to addiction—a chronic, relapsing disease characterized by compulsive drug seeking and use, which is accompanied by chemical and molecular changes in the brain. Some of these changes persist long after methamphetamine abuse is stopped. Reversal of some of the changes, however, may be observed after sustained periods of abstinence (e.g., more than 1 year).[4]

What Other Adverse Effects Does Methamphetamine Have on Health?

Taking even small amounts of methamphetamine can result in many of the same physical effects as those of other stimulants, such as cocaine or amphetamines, including increased wakefulness, increased physical activity, decreased appetite, increased respiration, rapid heart rate, irregular heartbeat, increased blood pressure, and hyperthermia.

Long-term methamphetamine abuse has many negative health consequences, including extreme weight loss, severe dental problems ("meth mouth"), anxiety, confusion, insomnia, mood disturbances, and violent behavior. Chronic methamphetamine abusers can also display a number of psychotic features, including paranoia, visual and auditory hallucinations, and delusions (for example, the sensation of insects crawling under the skin).

Transmission of HIV and hepatitis B and C can be consequences of methamphetamine abuse. The intoxicating effects of methamphetamine, regardless of how it is taken, can also alter judgment and inhibition and can lead people to engage in unsafe behaviors, including risky sexual behavior. Among abusers who inject the drug, HIV/ AIDS and other infectious diseases can be spread through contaminated needles, syringes, and other injection equipment that is used by more than one person. Methamphetamine abuse may also worsen the progression of HIV/AIDS and its consequences. Studies of methamphetamine abusers who are HIV-positive indicate that HIV causes greater neuronal injury and cognitive impairment for individuals in this group compared with HIV-positive people who do not use the drug.[5,6]

What Treatment Options Exist?

Currently, the most effective treatments for methamphetamine addiction are comprehensive cognitive-behavioral interventions. For example, the Matrix Model—a behavioral treatment approach that combines behavioral therapy, family education, individual counseling, 12-step support, drug testing, and encouragement for nondrug-related activities—has been shown to be effective in reducing methamphetamine abuse.[7] Contingency management interventions, which provide tangible incentives in exchange for engaging in treatment and maintaining abstinence, have also been shown to be effective.[8] There are no medications at this time approved to treat methamphetamine addiction; however, this is an active area of research for NIDA.

How Widespread Is Methamphetamine Abuse?

Monitoring the Future Survey[†]

Methamphetamine use among teens appears to have dropped significantly in recent years, according to data revealed by the 2009 Monitoring the Future survey. The number of high-school seniors reporting past-year[††] use is now only at 1.2 percent, which is the lowest since questions about methamphetamine were added to the survey in 1999; at that time, it was reported at 4.7 percent. Lifetime use among 8th-graders was reported at 1.6 percent in 2009, down significantly from 2.3 percent in 2008. In addition, the proportion of 10th-graders reporting that crystal methamphetamine was easy to obtain has dropped to 14 percent, down from 19.5 percent 5 years ago.

Use of Methamphetamine by Students 2009 Monitoring the Future Survey

	8th Grade	10th Grade	12th Grade
Lifetime	1.6%	2.8%	2.4%
Past Year	1.0%	1.6%	1.2%
Past Month	0.5%	0.6%	0.5%

National Survey on Drug Use and Health (NSDUH)[†††]

According to the 2008 National Survey on Drug Use and Health, the number of past-month methamphetamine users age 12 and older decreased by over half between 2006 and 2008. Current (past-month) users were numbered at 731,000 in 2006, 529,000 in 2007, and 314,000 in 2008. Significant declines from 2002 and 2008 also were noted for lifetime and past-year use in this age group.

From 2002 to 2008, past-month use of methamphetamine declined significantly among youths aged 12 to 17, from 0.3 percent to 0.1 percent, and young adults aged 18 to 25 also reported significant declines in past-month use, from 0.6 percent in 2002 to 0.2 percent in 2008.

Other Information Sources

For more information on the effects of methamphetamine abuse and addiction, visit www.drugabuse.gov/drugpages/methamphetamine.html.

To find publicly funded treatment facilities by state, visit www.findtreatment.samhsa.gov.

For street terms searchable by drug name, cost and quantities, drug trade, and drug use, visit www.whitehousedrugpolicy.gov/streetterms/default.asp.

Data Sources

† These data are from the 2009 Monitoring the Future survey, funded by the National Institute on Drug Abuse, National Institutes of Health, Department of Health and Human Services, and conducted annually by the University of Michigan's Institute for Social Research. The survey has tracked 12th-graders' illicit drug use and related attitudes since 1975; in 1991, 8th- and 10th-graders were added to the study. Questions about crystal methamphetamine were added to the 12th-grade and followup surveys in 1990, and questions about methamphetamine were added to the study for all three grades in 1999. The latest data are on line at www.drugabuse.gov.

†† "Lifetime" refers to use at least once during a respondent's lifetime. "Past year" refers to use at least once during the year preceding an individual's response to the survey. "Past month" refers to use at least once during the 30 days preceding an individual's response to the survey.

††† NSDUH (formerly known as the National Household Survey on Drug Abuse) is an annual survey of Americans aged 12 and older conducted by the Substance Abuse and Mental Health Services Administration, Department of Health and Human Services. This survey is available on line at www.samhsa.gov and can be ordered by phone from NIDA at 877–643–2644.

References

[1] Volkow ND, Chang L, Wang GJ, et al. Association of dopamine transporter reduction with psychomotor impairment in methamphetamine abusers. *Am J Psychiatry* 158(3):377–382, 2001.

[2] London ED, Simon SL, Berman SM, et al. Mood disturbances and regional cerebral metabolic abnormalities in recently abstinent methamphetamine abusers. *Arch Gen Psychiatry* 61(1):73–84, 2004.

[3] Thompson PM, Hayashi KM, Simon SL, et al. Structural abnormalities in the brains of human subjects who use methamphetamine. *J Neurosci* 24(26):6028–6036, 2004.

[4] Wang GJ, Volkow ND, Chang L, et al. Partial recovery of brain metabolism in methamphetamine abusers after protracted abstinence. *Am J Psychiatry* 161(2):242–248, 2004.

[5] Chang L, Ernst T, Speck O, Grob CS. Additive effects of HIV and chronic methamphetamine use on brain metabolite abnormalities. *Am J Psychiatry* 162(2):361–369, 2005.

[6] Rippeth JD, Heaton RK, Carey CL, et al. Methamphetamine dependence increases risk of neuropsychological impairment in HIV infected persons. *J Int Neuropsychol Soc* 10(1):1–14, 2004.

[7] Rawson RA, Marinelli-Casey P, Anglin MD, et al. A multi-site comparison of psychosocial approaches for the treatment of methamphetamine dependence. *Addiction* 99(6):708–717, 2004.

[8] Roll JM, Petry NM, Stitzer ML, et al. Contingency management for the treatment of methamphetamine use disorders. *Am J Psychiatry* 163(11):1993–1999, 2006.

INDEX

D

E

N

O

P

Q

R